THE NEW CITY

THE NEW CITY

How to Build Our Sustainable Urban Future

Dickson D. Despommier

Foreword by Mitchell Joachim

Columbia University Press   New York

Columbia University Press
Publishers Since 1893
New York   Chichester, West Sussex
cup.columbia.edu

Library of Congress Cataloging-in-Publication Data
Names: Despommier, Dickson D., author.
Title: The new city : how to build our sustainable urban future /
    Dickson D. Despommier.
Description: New York : Columbia University Press, [2023] |
    Includes index.
Identifiers: LCCN 2023009328 (print) | LCCN 2023009329 (ebook) |
    ISBN 9780231205504 (hardback) | ISBN 9780231556033 (ebook)
Subjects: LCSH: City planning. | Sustainable urban development.
Classification: LCC HT166 .D5463 2023 (print) | LCC HT166 (ebook) |
    DDC 307.1/216—dc23/eng/20230317
LC record available at https://lccn.loc.gov/2023009328
LC ebook record available at https://lccn.loc.gov/2023009329

Printed in the United States of America

Cover design: Milenda Nan Ok Lee
Cover art: James McNabb

I dedicate this book to all those whose lives were or still are focused on nurturing and advancing the development of the modern science of ecology: Charles Darwin, Rachel Carson, G. Evelyn Hutchinson, Aldo Leopold, Rene Dubos, Robert MacArthur, Theodosius Dobzhansky, Ruth Patrick, Eugene and Howard Odum, John Teal, Robert May, James Hansen, Gordon Eaton, David M. Gates, Arthur Tansley, Wallace Broeker, Edward O. Wilson, William Trager, James Lovelock, Lynn Margulis, John Cairns, Jr., Rita Colwell, Sylvia Earle, Roger Payne, David Attenborough, Cynthia Rosenzweig, Suzanne Simard, and Gene Likens. Without them, we would know very little about how rapid climate change came to be, and the devastating effects it is having on our planet.

Contents

Figures and Tables

TABLES

Foreword

Mitchell Joachim

nside *The New City: How to Build Our Sustainable Urban Future*, Dickson Despommier has given us a comprehensive, influential, and brilliantly compelling vision of how we can achieve a more sustainable future for ourselves and the planet. By taking a radically holistic approach to planning that focuses on the natural urban environment, Despommier shows us how we can judiciously address many of the most persistent environmental and social issues of our time, from climate dynamics to food insecurity to economic inequality.

As a pioneer in the field of vertical farming, Despommier brings a wealth of expertise to this demanding subject. He convincingly argues that we need to fundamentally rethink our approach to food production and distribution if we are going to generate truly green cities. But his vision extends far beyond just food—he investigates how we can use renewable energy, green infrastructure, and smart technology to transform our cities into pulsating, salubrious, and ecologically sound communities.

But perhaps what most sets Despommier's work apart is his unflinching larger-than-life optimism. Rather than simply pointing out the wicked challenges we face, he shows us that there are existent, tangible solutions within our grasp. By providing salient examples of flourishing initiatives from around the world, Despommier motivates us to believe that we can construct a better future for humanity.

Previously, Despommier has written on the subject of agriculture and urban farming, and his work has been featured in publications such as the *New York Times*, *National Geographic*, and *Scientific American*. He has also given countless lectures and vivid presentations, spreading awareness about the consequences of sustainable agriculture and encouraging others, including me, to take immediate action.

Thanks to Despommier's groundbreaking research and advocacy, vertical farming and eco-minded agriculture have become increasingly popular and have the potential to play a crucial role in feeding the earth's growing population while minimizing the negative effects of industrial farming. His colossal legacy will undoubtedly continue to inspire future generations to prioritize ecological thinking and make a net positive global effect.

The New City is a must-read for anyone who wants to create an innovative green society. Despommier's ideas are not just visionary—they are practical, achievable, and urgently needed. I have no doubt that this book will become a cornerstone of sustainability literature and will be read and studied for decades to come.

Preface

The world is mired in an environmental dilemma driven by rapid climate change (RCC). Humanity has created the lion's share of that problem by generating enormous quantities of atmospheric greenhouse gases (GHG), mostly through the combustion of fossil fuels. Cities annually produce nearly half of those GHGs. If RCC were an isolated phenomenon impacting a few million inhabitants, then it's likely that only those affected populations would be motivated enough to get involved. But this is not the case. RCC affects virtually all eight billion of us, as well as every other life form on Earth. If we want to survive into the next millennium, then we have no choice. We must all get involved to help resolve this problem.

Up until the late twentieth century, cities evolved guided largely by commercial interests and the politics of real estate. For development, location was and still is the prime concern. A project had to be economically sound and promise to increase the value of the property it occupied (i.e., show profitability), regardless of its purpose, or it did not get built. But most of the people who lived in that city and whose lives were directly affected by the often dangerous activity of construction and the attendant pollution and also by the kind of building being erected had no say in the matter. Only the developers, construction industry, and city politicians were ever consulted as to whether yet another commercial building was needed in their urban environment. Nor did any organization responsible for its "birth" consider the health of its future occupants or the people living in that building's neighborhood.

In the majority of large metropolises, only those who could afford to live in or adjacent to "action central," the urban nexus, where all the amenities of city life were located, benefited from living an urban lifestyle. City governments could have made all services available to everyone regardless of their address: outreach clinics and regional hospitals, quality schools, well-stocked supermarkets, and adequate numbers of police and fire stations. But just the opposite happened. Properties at a distance from the hub of commerce had a lower value. Yes, cheaper rents were common, but that meant there were fewer municipal services available, especially convenient access to health care. And the farther from the city center, the more likely it was that some pollution-generating utility or heavy manufacturing center was situated within those same disadvantaged neighborhoods. This arrangement remains essentially the same today.

But once in a great while, things do change. People tire of being taken advantage of. Some commit their lives to reforming city politics, and others work on technological solutions that benefit both humankind and wildlife. The latter efforts have led to (1) the development of efficient, affordable renewable-energy strategies; (2) carbon-capturing, recyclable construction materials with low carbon footprints; (3) cost-effective atmospheric water-harvesting methods; and (4) productive vertical farms situated within the city. Environmental justice issues are now front and center on many city council agendas. I call these four applications of technology the four pillars of sustainability. This gives hope that over the next hundred years, the world's cities will adopt some form of these technologies and pull themselves out of today's climate mess. Addressing this problem with human-centric urban design coupled with environmental stewardship is the only sensible way we can clear a path back to a more balanced, healthier life.

In the chapters that follow, I reflect first on the history of how we got to this point, and then I suggest what we can do to restore wildlife habitats and at the same time improve the way cities behave. Over the past hundred years, we have converted millions of acres of virgin forest into agricultural land, displacing whole ecosystems and forcing wildlife to adjust to living in damaged and fragmented habitats. This has led to the extinction of perhaps millions of species of flora and fauna. Most important, by deforestation we have eliminated a good portion of the carbon-recycling mechanisms that regulate the temperature of the Earth. The second half of my book details new, proven technologies and methods that address these issues. None of what I am about to describe is in the realm of science fiction. It is all out there waiting for a concerted effort on our part to tie these methods into a practical way of building that does no damage to the environment. Applying methods to the urban landscape that allow cities to obtain their own fresh water by harvesting it from the air, raise food crops in vertical farms inside the city limits, and use only renewable energy resources will allow humanity to finally live more harmoniously with the natural world rather than commandeering large portions of it for our own needs.

Acknowledgments

Books do not write themselves, nor does any author write one without help (sometimes a lot of help) from sympathetic friends and knowledgeable contacts willing to offer their time and creative input. My book benefited greatly from all those who volunteered their expertise and contributed useful comments, critical reviews, and sometimes candid opinions that on rare occasion managed to ruffle the feathers of the author. (But they were right, and I quickly got over it.) Regardless of the quality or quantity, all contributions were appreciated, and I wish to now thank them for keeping me focused on explaining why we need to reinvent the built environment.

Nothing would have happened had it not been for my book agent, Mel Parker. Mel helped negotiate my contract with Columbia University Press and has read so many drafts of the book that he's practically memorized it. His numerous observations and critical eye were crucial to me throughout the gestational period of *The New City*. Miranda Martin, my editor at Columbia University Press, was there for me throughout the creative process of turning what started out as a semirandom set of ideas into readable text arranged in a logical sequence. I thank Greg Kiss, architect and friend, for offering useful suggestions and much appreciated encouragement. I especially thank Scott Erdy, architect and teacher for his insights, detailed dissection of the book, and numerous useful suggestions for subtractions and additions to it. The book improved greatly as the result of his input. I thank William Schuster, forester extraordinaire, for leading me to critical insights regarding how a temperate-zone hardwood forest functions and survives. I thank Peter Walsh, dear friend, my long-lost brother number one, and master wordsmith, for his willingness to read through numerous drafts. With each read, Peter had a fresh set of comments that pushed the text to the next level. I thank Charles (Chuck) Knirsch, physician, globalist, teacher, and good friend for his patient and thoughtful critique of the book. His many insights into the origins of public health and the role of infectious diseases in shaping the history of urban life helped me improve the richness and depth of the manuscript. I thank Mitchell Joachim, creative genius and director of Terraform 1, architect, and out-of-the-box futurist, for sharing his unique views of the built environment and through numerous conversations explaining why and what we need to change to make it work. I thank Dr. Paul Brandt-Rolf, my former chairman of the Department of Environmental Health Sciences at Columbia University, for his thoughtful suggestions that made the text tighter and more readable. I thank Dr. Robert Fullilove, urbanist, scholar, artist, and long-lost brother number two, for his insights into what we need to do to make cities more equitable and ecologically responsible. I thank Laura Michele and Kevin Marshall for assisting me in obtaining some of the visuals, without which my book would not sparkle and shine. I thank Martha Helmers for contributing her graphics skills to generating the tables.

Then there are those that deserve special thanks. I begin with my colleague and long-time friend and teacher Constance (Connie) Halporn. Connie vigilantly sought out and obtained all permissions for the figures and made many useful formatting suggestions that significantly improved the overall feel of the book. Thank you, Herbert Sontz, dear friend and naughty limerick enthusiast (is there any other kind?), for helping me shape the book into what it finally became. Without your assistance, it would have taken a lot longer to piece together my ideas and thoughts. Your encouragement was inspirational. Finally, I thank my wife and best friend, Marlene Bloom, for her willingness to read and edit yet another one of my books. She did so with intelligence, grace, and skill. She also made many suggestions, all of which helped make the text more accessible.

Finally, I thank Dominic Casuli, my high school biology teacher, who saw something in me that, due to his unwavering encouragement, germinated and blossomed to define my life as a biologist.

THE NEW CITY

Introduction

Nature favors those organisms which leave the environment in better shape for their progeny to survive.

—JAMES LOVELOCK

HISTORY AND EVOLUTION OF CITIES

Cities are vibrant expressions of our need to interact and communicate. We are social animals, gregarious by nature. The urban environment is a product of our genetics, allowing us to share a common experience.

But today's cities are not sustainable in the long term. Their reliance on obtaining everything that enables them to function (e.g., food, water, energy) from outside the city limits makes them vulnerable when pandemics or wars break out, international trade agreements collapse, a power grid fails, or shortages of essential resources occur. Rapid climate change (RCC) is accelerating the collapse of a large number of global supply chains, particularly those related to food crops. Making sure that cities can access what they need to function endangers and often eliminates wildlife. What is more, many cities dispose of their liquid and solid wastes by dumping them back into the very same environments that they accessed for their needs.

The details have yet to be fully revealed by physical and cultural anthropologists, but the broad brushstrokes depicting our million-year journey from forest-dwelling primates to today are in full view for anyone who wishes to know our human origins. As we arose as a species on the plains of East Africa, we aggregated into small family units and then into larger and larger collections of kin. In time, we spread out over the entire face of the globe, except for Antarctica, and established large populations wherever we settled. During our migrations, we encountered and interbred with Neanderthal and Denisovan hominids. Agriculture arose, allowing our populations to increase. Cities sprang up next to farmland. Networks of manufacturing and consumption evolved as a by-product of our shift to a less nomadic, albeit more sustainable lifestyle.

Cities created the right environment for the emergence of civilization by the simple act of bringing people together. Written languages, mathematics, astronomy, industrialization, global commerce, organized religions, and a wealth of other cultural expressions grew out of our urban collectives. Today, the built environment is alive with change, motivated by our penchant to invent new solutions to old problems.

In the early days of cities, planning was often delegated by some religious-based or autocratic proclamation to a few individuals (enforcers) who then laid down the terms under which residents could carry out their daily lives. Cities grew with commerce as their overriding priority. The result, in many cases, was an unsafe environment. There is a long history of this negative aspect of life in the evolving urban landscape, with numerous examples of infectious diseases that took advantage of crowded places—plague, malaria, filariasis, typhoid fever, epidemic typhus, yaws, tuberculosis, syphilis, yellow fever, and smallpox, to name just a few of the more important ones. Noninfectious diseases also became a defining feature of the urban environment, with malnutrition, atherosclerosis, obesity, asthma, lung cancer, and a host of occupation-related health problems leading the way.

By the late 1700s, cities had matured and diversified. Most of them became essential destinations in a global network of trade centers. As a result, consumer demands increased exponentially.

As cities tried to keep pace with this unprecedented growth, the discovery of a powerful, concentrated, storable energy source eventually spawned the Industrial Revolution. Fossil fuels rapidly took over as the dominant choice for all energy-dependent devices. As a direct consequence, the climate began to change, eventually adversely affecting all living things. Over the next several hundred years, environmental conditions worsened even more as the use of fossil fuels dominated the energy market.

In 2021, nearly 57 percent of the world's population lived in cities. In another twenty years, that number will likely increase to 68 percent.[1] Cities are organic, expressing nearly all the characteristics of a living organism. Seen from the air at night, cities show off their beauty and geometry (see figs. 0.1, 0.2, 0.3). They gleam and sparkle. Cities are exquisite examples of living art. No two are alike, nor does any remain static. All are works in progress, collectively reflecting humanity's talent for creating new iterations of the urban experience.

But while each city is unique in character and structure, one thing has regrettably remained the same for all of them; they are plagued with health hazards and risks when compared to less crowded spaces.[2] As urbanization increases, this danger intensifies. Dust, exhaust fumes, debris, and inadequately handled liquid wastes contaminate the air, water, and soil. Heavy manufacturing still occurs within a large number of municipalities, adding more exotic pollutants to already overwhelmed waste-management systems. Industrial effluents include heavy metals and volatile and nonvolatile organic chemicals. Each year, lives lost to this toxic mix are measured in the millions. Even cities with state-of-the-art sanitation facilities have the same problems, only at reduced rates. Crowding is hazardous. Abandoned buildings attract vermin that can be dangerous to the homeless who often seek shelter there, presenting additional health risks. Traffic fatalities are among the top ten overall causes of death in many cities, annually killing some 1.3 million drivers and

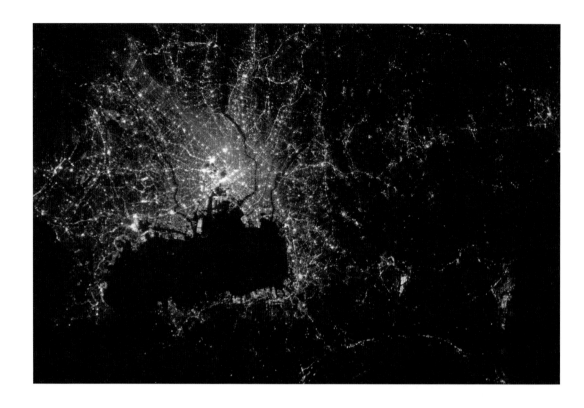

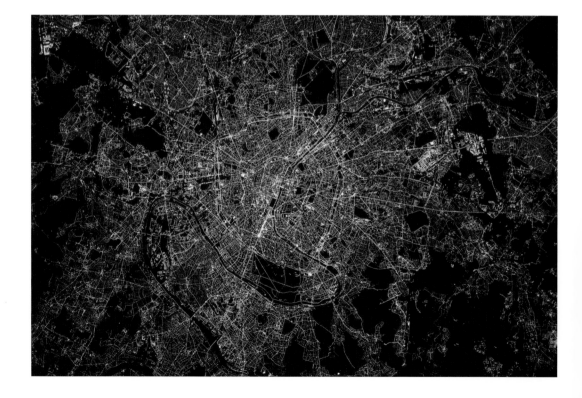

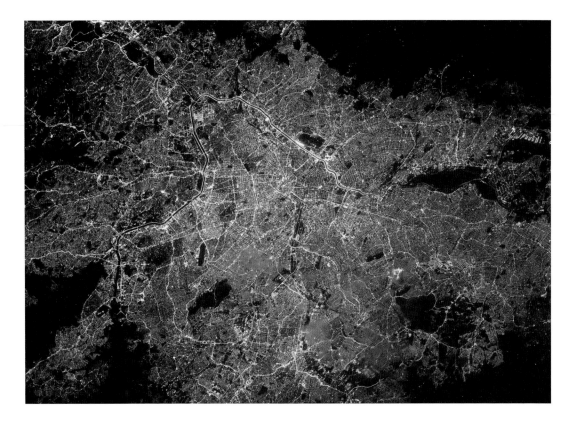

FIGURE 0.1 (*top, left*) Tokyo as seen at night from the International Space
Station. Courtesy NASA.

FIGURE 0.2 (*bottom, left*) Paris as seen at night from the International Space
Station. Courtesy NASA.

FIGURE 0.3 (*above*) São Paulo as seen at night from the International Space
Station. Courtesy NASA.

pedestrians.³ In less developed countries, the mortality rate for urban traffic-related accidents each year is much higher.⁴

In many cases, city infrastructure was purposefully designed to discriminate between the empowered and the poor. It's called environmental injustice. Why do so many urbanites continue to accept living under less-than-optimal conditions? Why are these roadblocks to a long, prosperous life not a thing of the past? After all, humanity continues to forge ahead, inventing and applying a cornucopia of science-derived applications that enable some to live much healthier lives than their ancestors. Antibiotics, gene therapies, monoclonal antibodies, new generations of antiviral drugs, mRNA-based vaccines, and robotic surgery have received accolades from all corners of the world, and deservedly so. Space travel is about to become routine, opening up new vacation destinations. Ground-based, ultra-high-speed hyperloop transport systems are under construction in many places. Regrettably, poverty remains endemic in all cities. Many modern advances will only be available to those fortunate enough to be able to afford them. The disenfranchised struggle each day just to stay alive. Inequality is an urban design flaw that must be corrected if we are all to have a chance at a better, healthier life.

We have lived in cities for thousands of years, so one would think that by now we would have evolved systems of representative governance that respect the individual and encourage innovation and creativity. Urban landscapes should have become healthier places for all to enjoy and maintain for generations to come. At the very least, they should be places that by design pose no threat to our longevity and well-being.

But as everyone who lives in one knows, that is not how cities grew and developed. None of the desirable qualities of urban life were embedded into their master plan. Commerce was and is the driving force in the evolution of the built environment. Densely populated areas have massive problems with public health, environmental justice, and the exclusion of nature. Even those cities severely altered by natural forces such as earthquakes and floods or human-caused destruction by wars, fire, and urban blight arise again essentially the same as they were before those catastrophic events.

It is unfair to condemn all metropolises for enforcing exclusionary policies since many do not, but despite any city's best intentions, especially mega-metropolises, they remain unsafe places, especially to those in the lower income brackets. Urban planners and developers follow patterns of design established by tradition and proven functionality. No city has ever been created so that all residents' lives are considered equally deserving of the benefits of living there. Nor has any been designed with improving the health and safety of its residents as its main goal. None have looked to nature as a guide, emulating the behavior of some sustainable animal or plant community. That is about to change.

There has never been a shortage of ideas aimed at improving city life—from the pharaohs of ancient Egypt to the engineers and planners in China, Peru, and imperial Rome. That tradition continued into the twentieth century with visionaries like Le Corbusier, Buckminster Fuller, Frank Lloyd Wright, and Frederick Law Olmsted Jr. A few of today's leading figures who offer new perspectives on the built environment include Saskia Sassen, Edward Glaeser, Janette Sadik-Khan, William McDonough, and Allan Jacobs. Organizations that focus their attention on urban issues include Smart Cities Council, IBM, Siemens, and the Massachusetts Institute of Technology. The smart city concept is theoretically a good one, but even with the large number of groups claiming that phrase in their title, it remains a diffuse, uncoordinated, and nonintegrated collection of ideas.

Many issues apply to all cities, not just the ones that cannot afford the technologies that create more secure lifestyles for their residents. The United Nations has issued what amounts to an indictment of the environmental crisis we face in coming to grips with the behavior of cities. The UN states, "Cities are major contributors to climate change. According to UN Habitat, cities consume 78 percent of the world's energy and produce more than

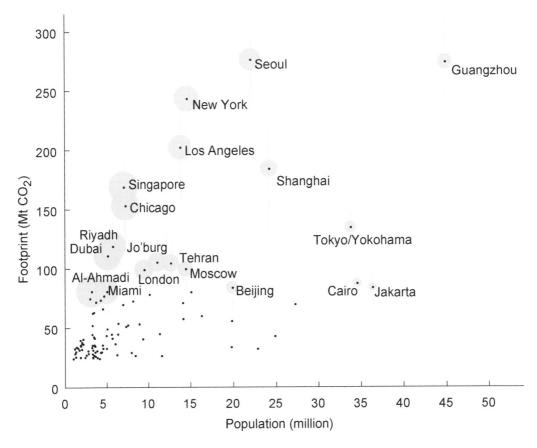

FIGURE 0.4 Carbon footprint of selected cities throughout the world. *Source*: Daniel Moran et al., "Carbon Footprints of 13,000 Cities," *Environmental Research Letters* 13, no. 6 (June 2018): 064041.

60 percent of greenhouse gas emissions. Yet, they account for less than 2 percent of the Earth's surface" (see fig. 0.4).[5] This is reason enough to demand radical changes in the way cities function.

RAPID CLIMATE CHANGE

Many people are confused about the difference between climate and weather. Weather varies, often daily, but its general quality—dry, wet, hot, or cold—follows the same direction as the climate. The difference between them is the time it takes for each to show

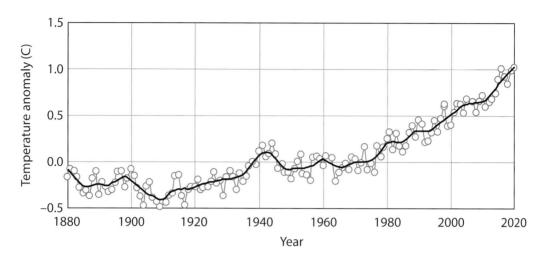

FIGURE 0.5 Rise in average annual temperature of the Earth. Courtesy NASA.

a noticeable trend that differs from the present situation. Our climate is also always changing, but it is rarely a rapid process and thus not easily observed and measured, at least not up until the present.

From a variety of science-based studies carried out over the past twenty years, there is a wide consensus that the relatively straight-line direction that our climate has taken over the previous 100,000 years of history abruptly changed with the beginning of the Industrial Revolution (see fig. 0.5). The surface temperature of the world's oceans regulates the Earth's climate, and it has been rising at an accelerating rate since the nineteenth century (see fig. 0.6).

One of the main drivers of Earth's climate is the Gulf Stream, emerging from the Gulf of Mexico at the tip of Florida and flowing northward along the eastern coast of the United States until it reaches Cape Hatteras, North Carolina. There it is deflected farther offshore into the Atlantic Ocean (see fig. 0.7). It continues north until it encounters Newfoundland, where it divides into two parts. One branch continues across the Atlantic, warming the western coast of the British Isles, Ireland, Norway, and the

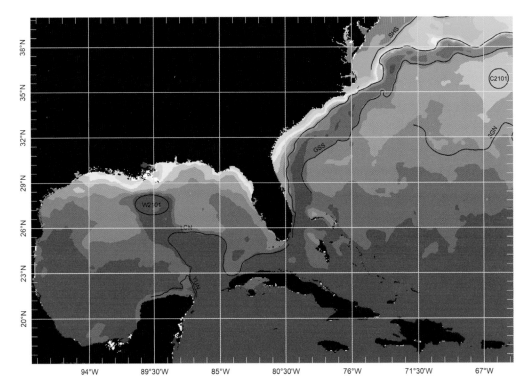

FIGURE 0.6 Average annual ocean temperature in the Gulf of Mexico has been rising each year since measurements were first recorded. Courtesy NOAA.

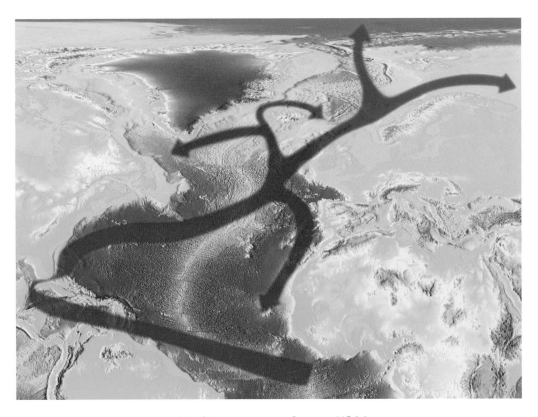

FIGURE 0.7 Schematic diagram of Gulf Stream currents. Courtesy NOAA.

southern coast of Iceland. The other branch flows further north and butts up against the southern coast of Greenland.

Every winter when seawater freezes solid and forms the Arctic ice sheet, it "squeezes out" the salt and creates a hypersaline solution. Because it is denser than normal seawater, that solution sinks, helping to generate a powerful downward current of cold water. The other "engine" that contributes to that current is created when the northern branch of the Gulf Stream encounters the colder air of the Arctic. It becomes denser as it cools, and it, too, begins to sink.

This powerful current, known as the grand ocean conveyor belt (see fig. 0.8), was first described by Wallace Broecker and colleagues at the Lamont-Doherty Earth Observatory at Columbia University. Oceanographers refer to it as the Atlantic meridional

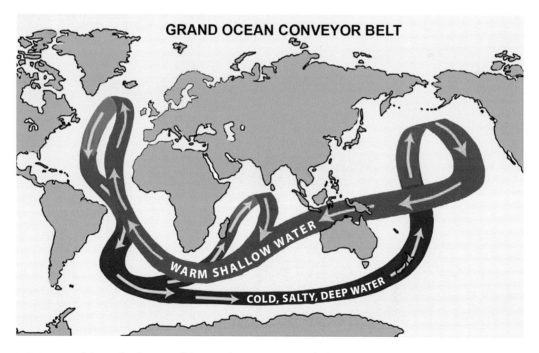

FIGURE 0.8 Schematic diagram of the grand ocean conveyor belt. Courtesy Lamont-Doherty Earth Observatory.

overturning circulation or AMOC. It encircles most of the Earth and is constantly in motion, completing a cycle roughly every thousand years. It is responsible for sea surface temperatures on a global scale, which in turn affect the air temperature. For example, in the Atlantic Ocean, cold water rises to meet the southeastern coast of Africa, creating the nutrient-rich Agulhas Current and nurturing all the aquatic plants and marine animals found there. The climate along that same region is temperate trending on the cool side.

There is a strong correlation between RCC and the circulation of the world's oceans. The geological record is rich with abrupt changes caused by an irregularity in the grand ocean conveyor belt. Weakening the AMOC would affect virtually every life form on the planet. Because the ocean's circulation is caused by the presence of hypersaline water and a cooling Gulf Stream in the Arctic, a change in either process would affect the strength of the current. One way to slow down the AMOC is by adding more fresh water to the Arctic Ocean, diluting it, and making it less dense. The ice sheet in Greenland consists of fresh water and averages over a mile in thickness. It has been melting, pouring millions of gallons of fresh water into the sea at an unprecedented rate over the past twenty years.

The other mechanism for generating the current is the freezing of saltwater when the Arctic ice sheet develops each winter. Satellite data clearly show the dramatic effect of RCC on the annual freezing of the Arctic Ocean. The ice sheet has become thinner and smaller in area each year since 1960. In 2020, the ice sheet was at its second-lowest recorded level (see fig. 0.9).[6] The unavoidable conclusion is that the grand ocean conveyor belt is threatening to slow down, which will lead to annual increases and severity of floods, droughts, hurricanes, and other adverse weather-related phenomena.

Climate change occurs over years to decades, not days or months. Yet many people question whether we are responsible for altering the climate. Many of those "climate deniers" cite

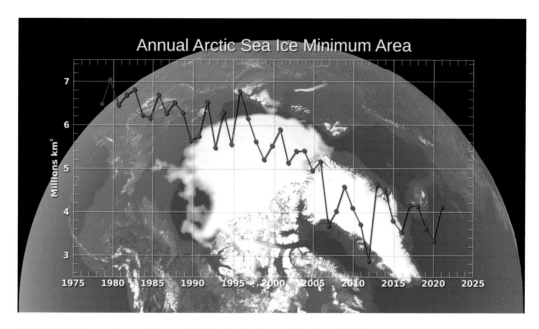

FIGURE 0.9 Annual loss of ice pack in the Arctic Ocean. Courtesy NASA.

personal observations as the basis for their skepticism. Others say that it's all part of the natural cycle. The National Aeronautics and Space Administration (NASA) and the National Oceanographic and Atmospheric Administration (NOAA) provide indisputable evidence that RCC is real and that human activities are the cause of most of it. These two governmental agencies collect data and present them to the public, allowing everyone to evaluate the information independently. They do offer opinions and analysis, but science is the basis for their carefully thought out, data-based interpretations. NASA has presented a set of facts about the realities of RCC.[7] Its shortlist of the reasons we know for sure that the planet is changing faster than we can adjust to is presented here:

- Global temperatures have risen 2.2 °F (1.8 °C) since the nineteenth century.
- The ocean is warming, rising 0.6 °F (0.33 °C) since 1969.

- The Greenland and Antarctica ice sheets are melting.
- Globally, nearly every glacier is in retreat mode.
- Snow cover is less, and snowmelt is more rapid in the Northern Hemisphere over the last five decades.
- The world's oceans rose roughly eight inches within the last one hundred years.
- The Arctic ice sheet has been greatly reduced over the last several decades.
- The number of record high temperature events in the United States has been increasing, and the number of record low temperature events has been decreasing, since 1950. The United States has also witnessed increasing numbers of intense rainfall events.
- Since the beginning of the Industrial Revolution, the acidity of surface ocean waters has increased by about 30 percent.

Many other alterations in our world corroborate and strengthen the conclusions of climate research in disparate fields within the natural sciences. All of the large-scale observations listed here are having adverse effects on every aspect of how ecosystems behave.

One reliable biological indicator of how RCC affects wildlife is its detrimental effect on shallow-water coral reef biology.[8] The bleaching of coral (see fig. 0.10) is an observable indicator of the negative effects of ocean warming.

NOAA estimates, "In 2016, heat stress encompassed 51 percent of coral reefs globally and was extremely severe—the first mass bleaching (85 percent bleached) of the northern and far-northern Great Barrier Reef killed 29 percent of the reef's shallow-water corals."[9]

In addition to warming, the ocean is also experiencing an increase in acidity.[10] Before the 1950s, the pH (i.e., the amount of dissolved acid) of the oceans measured 8.2. As the pH decreases in number value, the acidity increases. In 2021, the pH had dropped to 8.1. Many areas of the world's oceans show signs of greater acidification, with some regions reaching levels approaching

FIGURE 0.10 Healthy coral reef (*left*); bleached coral reef caused by a warming ocean (*right*). Courtesy NOAA.

pH 8.0–7.9. This acidification is entirely caused by the increase in anthropogenic carbon dioxide (CO_2) in our atmosphere. Since deforestation has short-circuited Earth's carbon sequestration mechanism, much of the CO_2 that used to be taken up by trees and wild grasses now dissolves into the oceans, where it forms carbonic acid. A lower pH is bad news for shellfish of all kinds and for shallow-water corals. These life forms produce calcium carbonate as a by-product of their metabolism and use it as a building material for shells and the solid portion of reefs. Exposing calcium carbonate to an acid environment below pH 7.6 (see fig. 0.11) causes the calcium to disassociate from the carbonate moiety, and the compound literally falls apart.

Such is already the case for Tacoma, Washington. In 2010, the pH of the water from Puget Sound in that area had already dropped to 7.9, and various species of mollusks collected during the following years from that locale clearly showed signs of aberrant shell formation.[11]

There are more markers of the effects of climate change. The world's freshwater lakes are warming up, resulting in a reduction in the amount of dissolved oxygen that their waters can hold,

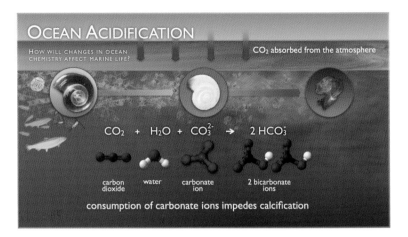

FIGURE 0.11 Schematic diagram of the negative effects of ocean acidification on shellfish. Courtesy NOAA.

threatening to disrupt the balance of energy flow among the complex networks of life that live there.[12]

The annual increase in the number and intensity of forest fires is another example of how our changing climate has affected both the natural world and the built environment. These events have measurable long-term undesirable effects on biodiversity and reduce the ability of forests to sequester carbon from the atmosphere. In addition, there is a human health risk from large-scale forest fires. Inhaling the small-particle fraction of smoke exacerbates and intensifies asthma attacks and related respiratory illnesses, such as chronic obstructive pulmonary disease.

The unbalancing of the ecosystems from climate change is also revealed in the changed cycles of other animal species. The mountain pine beetle (*Dendrochtonus ponderosae*) is an insect pest of western pine forests, mainly in British Columbia. The insect's preferred host tree is the lodgepole pine. The adult beetle lives just under the bark of that tree, where it lays its eggs. The eggs hatch, and the larvae begin consuming living cells from the phloem, an essential cellular component that distributes nutrients to all parts of the tree. As winter approaches, the larvae go

into diapause, suspending their development to adulthood until
the following spring. If winter temperatures fluctuate between
–13 °F and –31 °F, the larvae die, and the tree survives. Because
of rising temperatures, this beetle species has been extending
its range, moving steadily northeastward (see fig. 0.12). That is
because over the last ten years, winters in the Pacific Northwest,
especially in British Columbia and western Alberta, have been
mild, and the beetles have been able to survive and complete
their life cycle.

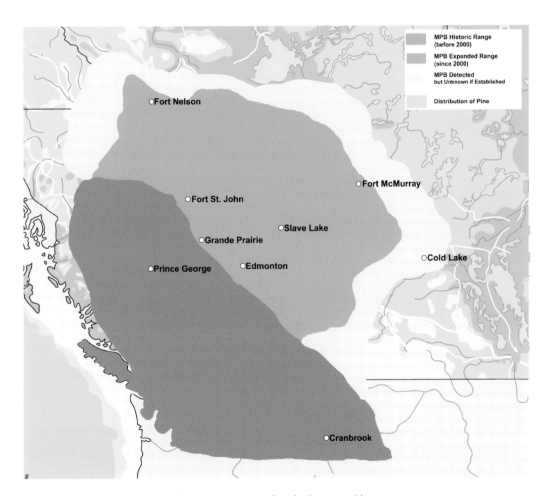

FIGURE 0.12 Extended range of the mountain pine beetle due to rapid
climate change. Courtesy Government of Canada.

The mountain pine beetle has killed extensive swaths of forest. The fate of large stands of dead trees is predictable—more devastating forest fires. Environmental scientists point to RCC as the underlying cause that triggered the expansion of the range of the beetle's population.[13]

Many birds time their annual migrations to coincide with the arrival of spring and the leafing out of trees. The new tender leaves provide food for many varieties of insects, and they, in turn, provide food for the birds. With warming trends accelerating, the trees leaf out earlier, but most bird species arrive around the same time they always have. By late spring many insect species have already taken advantage of the early season leaf cover to grow and develop into flying adults and are now better able to evade capture. The absence of caterpillars and other larval insect forms limits the food parents can supply for their chicks. Mortality rates among many songbirds are increasing because the birds were too late to "catch the early worm."[14]

Some groups of swallows have adapted to early spring and now lay their eggs sooner. These acrobatic birds are adept at catching their meals on the fly, so the presence of adult insects is not the problem. However, adverse early weather events, such as a cold spell, a frequent annual occurrence, can result in a higher mortality rate among the swallow hatchlings.[15]

Not everything is affected by the rise in temperature of Earth's atmosphere. The timing of the emergence cycle of the seventeen-year cicadas (*Magicicada septendecim*) has been a well-known phenomenon of nature for centuries. That insect is remarkably predictable in terms of its periodicity. So far, RCC has not significantly affected its emergence schedule. But scientists who study its biology predict that as warming increases, the cicada will eventually adjust and begin morphing into the adult stage earlier than its current cycle time.[16]

Without a concerted global effort to reduce the level of greenhouse gasses emitted by our many activities, there is little hope of slowing down the rate at which our climate changes.

URBAN DESIGN

It is worth reemphasizing that cities occupy just 2 percent of the world's landmass but generate over 60 percent of the greenhouse gases that contribute to RCC. The extensive use of concrete and steel in urban construction is a major contributor to cities' large carbon footprints.[17]. As glaciers melt and sea levels rise, eventually hundreds of millions of coastal dwellers may have to move inland. The way we build and manage cities must now be guided by an overarching principle—do no harm to the surrounding landscape.

Making cities independent of the surrounding landscape would go a long way in addressing climate change. In 2015, I began teaching an undergraduate course entitled Ecology for Designers at Fordham University in New York City. Using the same didactic approach that I implemented for the vertical farm concept, we investigated technologies that applied to urban design and whose intent was to get cities off the centralized municipal grids for energy, food, water, and waste management. I asked the class to imagine a city of a hundred thousand people that behaved autonomously. We called it Fordhamopolis. It has a website (fordhamopolis.com), and I encourage you to visit it. The students enthusiastically bought into the premise of our theoretical construct and came up with many alternative ways to achieve the goal. We redesigned the way cities accessed their resources and eliminated single-source grids. Temperate-zone trees served as our inspiration from nature. We incorporated many of their biological characteristics into our urban design. We proposed that each building should be able to sequester carbon, harvest rainwater, grow its own food, and generate its own energy from solar radiation. We named this integrated approach *achieving gridlack*.

Surprisingly, the technologies incorporated into all the city's buildings that matched up best with those tree functions were not the least bit experimental or untested; all were up and running.

A few of them needed more tweaking in research and development to become robust enough to be included in the infrastructure of our new city, but all of them are commercially available.

The question was why aren't they being used more extensively to repurpose the built environment into something that is not a burden on the rural countryside? No answer meant that business as usual would prevail until things got much worse. I began writing. Perhaps I can coax a few deep pockets to take up this challenge as others have already done for the concept of vertical farming.

INSPIRATION FROM INTACT BIODIVERSE TEMPERATE-ZONE FORESTS

Forests capture atmospheric CO_2 as trees use it to synthesize cellulose, one of the primary molecular building blocks of wood. Capturing and storing carbon (the carbon sequestration process) helps regulate Earth's climate. Trees provide that service free of charge, but over the last ten thousand years, we have cleared away some three trillion of them, mainly to make room for farmland.[18] Bringing the atmospheric portion of the carbon cycle back to near its pre–Industrial Revolution concentration (280 ppm of CO_2) will require restoring a good portion of Earth's forests, as well as switching to renewable energy resources.

Other sources of excess CO_2 in our atmosphere come from industrial processes associated with the acquisition, production, and use of today's building materials: steel, concrete, and glass. In contrast, using wood as the construction material for all the future structures of a city is in keeping with a *do not harm* bioethical policy. Forests will eventually regenerate, replacing the missing trees that we selectively harvest for rebuilding our cities. Cross-laminated timber (CLT), a newly developed engineered wood product for building construction, constitutes a potentially significant urban carbon storage depot.

In too many places throughout the world, access to safe drinking water is not available (see fig. 0.13). That is because global

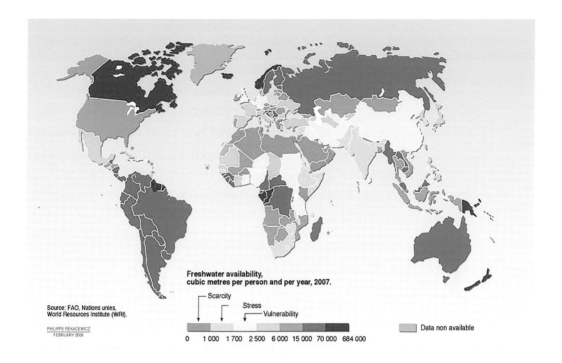

Freshwater availability,
cubic metres per person and per year, 2007.

Scarcity
Stress
Vulnerability

Source: FAO, Nations unies,
World Resources Institute (WRI).

PHILIPPE REXACEWICZ
FEBRUARY 2008

0   1 000   1 700   2 500   6 000   15 000   70 000   684 000

Data non available

FIGURE 0.13 Global distribution of availability of freshwater. Courtesy
WHO/UNICEF Joint Monitoring Programme for Water Supply, Sanitation,
and Hygiene.

agricultural practices continue to consume and despoil over
70 percent of the available fresh water annually with agrochemicals such as pesticides, herbicides, and fertilizers. The new city
will acquire its water the same way Bermuda gets theirs: from the
air. More on that later.

Like all other green plants, trees capture the energy of sunlight
through their leaves via the complex chemistry of photosynthesis. This not only fuels the growth of the tree but also supplies
the energy needed for the production of an astonishing assort-
ment of tree products, many of which are essential food items for
the animals that live in balance with the forest and, among other
things, help to distribute seeds into the surrounding environment. This safeguards the survival of the entire forest ecosystem.

These two features, interdependency and resiliency, are worth emulating in the new city.

The seamless integration of carbon storage, water harvesting, food production, and energy generation, coupled with their ability to share these resources, sustains the long-term survival of temperate-zone forests. Inspired by nature, we can permanently incorporate these same attributes into the fabric of the urban landscape, and in doing so, no city dweller will get left out. We can create a world with no shortage of food or fresh water, no substandard housing, and no blackouts—a net-zero city.

Lots of things thought to be impossible have come about over the last twenty years thanks to a series of innovations that have dramatically improved the lives of billions of people. Affordable cell phones, the internet, and rapid public transportation systems, to name but a few, have shown the way to a more equitable future. Hope and positivity: we need both, together with a willingness to embrace technologies that emulate natural processes and then put them to work for everyone's benefit.

Forests have persevered over millions of years, and along the way, through the process of natural selection, they acquired biologically redundant systems that ensured their long-term survival. Cooperation—sharing essential resources among the trees of a given biome—is one of the significant features of that long process of trial and error. It allows trees to survive and thrive, regardless of the often-harsh conditions of a constantly changing environment. Cities need to incorporate this aspect into their master plan for resilience if they, too, are to thrive into the next millennium.

Consider the following: forests have survived four mass extinction events, including the meteor six miles in diameter that wiped out more than 75 percent of all life on Earth sixty-six million years ago. Recovery from that global apocalypse in most places throughout the Northern Hemisphere was comparatively rapid, taking just over a million years. During the next six million years, the rest of the world's forested ecosystems gradually returned to their premeteor level of density but were composed of entirely new species of trees. Resilient nature persisted, and there were once again verdant woodlands overflowing with life in many parts of the world. It's all quite remarkable when one considers how bleak and desolate the Earth must have looked soon after that large piece of rock slammed into the Yucatán Peninsula.

Evolution is life's strategy for creating a shared genetic memory bank. We have figured out how, using scientific research, to open the safe where that information is stored. Over the last fifty years, ecologists throughout the world have learned how trees in temperate-zone forests collaborate as communities to provide for their longevity. A more detailed look into how forests function will give readers a better appreciation for why this ecological example is the best one to emulate for our new city.

Gene Likens studies mechanisms that underlie forest resilience following catastrophic events, such as clear-cutting (see fig. 0.14) an entire watershed (see fig. 0.15). From their original studies in the 1960s and early 1970s on the Hubbard Brook watershed 2 in New Hampshire, Likens and his colleagues determined how a forest recovers when a portion of it becomes damaged by clear-cutting and is then left alone to repair itself. The fast-growing understory (shrubs and bushes) played a major role in holding in the soil, water, and nutrients (particularly nitrogen) following that event until the slower-growing trees finally grew taller and resumed the maintenance of the forest. The entire process took over twenty years. Throughout their studies, which continue to this day, they collected water and soil samples to track the repair process. Their investigations resulted in numerous seminal publications. Many other forestry research data sets supported their conclusions. Strong scientific evidence now reveals the underlying mechanisms forests employ to make them resilient, renewable resources.[19]

Another forestry pioneer, Suzanne Simard, working in Oregon in the 1990s, single-handedly discovered many of the basic biological features responsible for the long-term maintenance

FIGURE 0.14 A portion of clear-cut forest. USGS.

FIGURE 0.15 An entire clear-cut watershed. ©2013 Walter Siegmund

of old-growth forests. Collecting data over four years during her doctoral research, she discovered that trees actually "talk" to each other by employing complex chemical signaling networks. She showed that some tree-to-tree signals are transmitted through the air, while others are mediated by connections between soil fungi and tree roots(see fig. 0.16).

FIGURE 0.16 Cross-section of soil and seedling showing connections between soil fungi and tree roots. Courtesy NC State.

Following the publication of her groundbreaking findings, a chorus of experts expressed disbelief in her conclusions. However, Simard's trust in the methodological soundness of her experiments did not waver. Eventually, she won the day but only after other less skeptical researchers replicated her experiments and found the same results in other patches of forest. Most of the establishment foresters finally embraced her findings. Simard's studies continue to inspire a new generation of temperate-zone forest ecologists, yielding more insights into the biology of trees, with emphasis on their survival mechanisms. In her seminal book, *Finding the Mother Tree*, Simard elegantly describes how certain older trees manage to "stay in touch" with their seedlings through mycorrhizal (fungal) networks in the soil. Her main conclusion from years of research is that to guarantee the sustainability of a healthy forest the older trees must be left alone. This helps maintain biodiversity. Simard states, "Somehow with my Latin squares and factorial designs, my isotopes and mass spectrometers and scintillation counters, and my training to consider only sharp lines of statistically significant differences, I have come full circle to stumble onto some of the Indigenous ideals: Diversity matters."[20]

The success of temperate-zone forests in terrestrial environments is the result of eons of environmental exposure and adaptations to changing climates. Geographic isolation due to plate tectonics—the slow, steady movement of continents—also contributed significantly to creating new species of plants and animals. Humidity, soil moisture, and elevation exerted their influence in shaping how life evolved for local environments. The Earth's temperature was more tropical as the age of trees began, about 380 million years ago. The weather was warm, and the climate changed slowly since the surface of the Earth was still in cooling down mode. The first trees were simple compared to modern-day species. There were only a few varieties, and all had very similar characteristics. None produced flowers or seeds. Despite these limitations, it was possible for forest ecosystems to become established.

By the Carboniferous Period (around 354 million years ago), the progenitors of the pines (gymnosperms) had arisen and were flourishing—the first trees to produce cones containing seeds. The conifers evolved into numerous species, occupied many new habitats, and established robust, abundant populations. Many descendants of the original pine tree family are still thriving. Currently, coniferous evergreens comprise about eight hundred tree species.

Flowering plants (angiosperms) developed later during the Cretaceous Period along with an explosion in insect diversity. Beetles and ants were some of the world's first pollinators. As of 2020, according to the Botanic Gardens Conservation International organization, there were over 369,000 known species of flowering plants, including 59,200 tree species that produce seeds from flowers.

Through their intimate connections with the tree roots, certain mycorrhizal fungal species allow many of the trees in each region to share water and nutrients. These fungi accomplish this by absorbing essential elements and water from the aqueous portion of the surrounding soil matrix and then sharing them with their hosts. In return, the trees supply the fungi with energy-rich sugars and other products of their metabolism that the fungi cannot synthesize; mycorrhizal fungi, and all other fungal species, cannot carry out photosynthesis. This mutualistic, symbiotic relationship between trees and their fungal associates is worthy of translating into the built environment. Creating buildings that cooperate with one another by sharing essential resources is one of the main goals of our new city.

As stated earlier, ecosystems represent the best examples of how assemblages of plants, animals, and their associated microbial communities create and conserve energy and maintain nutrient recycling networks that enable all the life forms within a given geographic region to remain in balance. This essential design feature is the basis for the four pillars of sustainability. Each building in the new city will behave similarly to the ways a

tree behaves in a balanced ecosystem with respect to how it generates energy, produces its own food, harvests water from the air,
and sequesters carbon.

A FLAWED EXAMPLE: AGGRESSIVELY COMPETITIVE TROPICAL FORESTS

Unlike temperate-zone forests, many tree species in the tropics
aggressively compete with one another for sunlight and nutrients,
exhibiting no evidence of altruistic communal behavior. While
their root systems are also connected by mycorrhiza, tropical species of underground fungal networks do not appear to function
by sharing essential resources.[21] This difference between tropical
and temperate-zone forests may have arisen as the Earth cooled
and temperate climate zones became the norm, forcing the trees
in those areas to set aside some of their energy reserves, allowing them and their root fungal associates to survive together
throughout the winter without the need for photosynthesis. For
tropical trees that keep their leaves year-round, photosynthesis is
continuous.

Some tropical tree species have a parasitic relationship with
others. A good example is the aptly named strangler fig, whose
seeds are deposited on the branches of an unsuspecting tree
along with the feces of a diverse group of fig-eating animals, such
as birds, bats, and nonhuman primates. The seeds only germi-
nate once the animals excrete them. The parasitic seedlings send
down aerial roots to the forest floor, where they burrow into the
ground and spread out around the helpless host tree. The strangler fig outcompetes its victim for nutrients and water and eventually kills its host (see fig. 0.17).

Tropical forests and their temperate-zone cousins both also
harvest and store rainwater. Tropical trees use this water through
evapotranspiration, common to all higher green plants, to initiate rain events that typically occur daily during the wet season.
Most of the precipitation they generate by this process is absorbed

FIGURE 0.17 Strangler fig tree slowly killing its host tree. By Poyt448 Peter Woodard

back through their shallow root systems, maintaining the cycle of rain production. Rain leaches minerals from the shallow tropical soils, making them available for the trees. Rain also washes minerals from the surface of leaves, allowing each tree to reabsorb a portion of its pool of essential nutrients and encourage new branches to sprout on the lower parts of its trunk. Perhaps in this regard, tropical trees empower the collective forest to benefit from massive water releases into the atmosphere each day. But such behavior is less valuable to planners seeking to emulate a natural process in creating a new urban environment. During the dry period, in seasonally arid tropical forests, trees rely heavily on their stored water, enabling them to make it through to the next rainy season. But even here, the difference between tropical and temperate-zone trees is apparent. No tropical tree species shares its stored water or pools of nutrients with its adjacent tree neighbors. In the tropical rainforests, it's every tree for itself.[22]

1

Pillar One
Carbon Storage

The twisted tree lives out its life.
The straight tree ends up as a board.
—CHINESE PROVERB

Besides restoring the world's fragmented forests to their carrying capacity of six trillion trees, there are few other practical solutions to rebalancing the Earth's carbon budget. The Trillion Tree Campaign by Plant for the Planet[1] is attempting to restore a portion of the damaged forests through community-funded and -staffed initiatives. I commend the organization for its efforts and hope very much that it succeeds, but one problem remains: Where will those trees be planted if farmland continues to increase its footprint as our population grows even larger? It's time to find other ways to produce our food. That is why urban agriculture is included in my four pillars of sustainability and reflects one of the main functions of trees—supplying forest-dwelling wildlife with food resources.

If planting three trillion trees seems like a daunting task, it is. I recommend taking inspiration from a remarkable Pulitzer Prize–winning short story by Jean Giono, "The Man Who Planted Trees." Its plot is simple enough: A middle-aged man living in the south of France is committed to planting one hundred tree seeds per day for the rest of his life to try to bring his desecrated, barren part of Europe back to a more livable environment—not just for him but also for the rest of the wildlife that abandoned their

33

destroyed habitats. The trees had all been cut down, turned into charcoal, and sold. No more trees meant no more wildlife to support the needs of the local communities. The people who caused this problem had long ago moved out of the area. Giono's story is as much about the power of one determined person and his passion for life as it is about restoring a damaged environment. The story is familiar to all of us who have wondered what happened to all the wild spaces that used to be part of our childhood.

In the story, Elzéard Bouffier lives in Provence. Every day, he treks across the countryside, stopping every now and then to plant a seed or two—acorns at first then other tree species native to the general area. Eventually, a forest is created out of nothing more than the relentless determination of a single human being. Over a forty-year period, Bouffier single-handedly plants 1,460,000 trees. The environment improves, the rivers and streams again flow at full volume, and wildlife reclaims their restored habitats. Bouffier's forest is now self-propagating by annual natural seed production.

Could a simple restoration project such as this become an essential part of the solution to our need to replace the three trillion missing trees? Planting one hundred seeds should take less than two hours if one seed is planted every five steps. In a day, planting four hundred seeds seems like a reasonable number assuming that half of them survive to maturity. So how many of us might it take to get the job done? A person working for one year could start growing 73,000 trees. A million people could then produce as many as seventy-three billion trees, but planting a trillion trees would take a very long time. It would be better to let the forests naturally reseed themselves. All we need to do is leave them alone, and they will take care of the rest, and it will happen in a much shorter time frame than we could ever hope to achieve by planting them ourselves. For example, a mature maple tree produces over ninety thousand seeds per year, while birch trees can produce over a million seeds. Douglas fir trees over their life span of a thousand years can potentially produce ten million seeds. Another reason to let the forests do the job is that in a climax (mature) hardwood forest, biodiversity is maintained. Giono's book reinforces this principle and offers an excellent description of the role trees play in helping to foster balance, biodiversity, and sustainability in an undisturbed environment.

CARBON FARMING

The Intergovernmental Panel on Climate Change (IPCC), in its 2015 recommendations for mitigation of rapid climate change (RCC),[2] stated that increasing the rate of carbon capture from the atmosphere is our best hope for remediation of the environment. The IPCC suggests, among other things, that farmers should switch to trees as their crop of choice and become "carbon" farmers. I wholeheartedly endorse their recommendation as long as the forests they create are biodiverse and not monoculture tree plantations. Since farming of all kinds has dominated the landscape, why not encourage growers to convert to a set of plants renowned for their ability to capture and store carbon in the form of a readily available and renewable building material—wood?

BUILDING CITIES OUT OF WOOD

One of the most revered and well-known timber frame structures, the Buddhist temple Hōryūji, is located in Ikaruga, Japan (see fig. 1.1). Buddhists built the structure from trees 1,300 years ago. It is a remarkable example of wood as a durable construction material. Other wooden structures, such as the eight hundred buildings of the Forbidden City in Beijing, completed in 1420, still stand tall. Examples of antique wooden buildings, primarily churches built four to five hundred years ago, are scattered throughout Europe.

Timber-frame construction is still in fashion, from the Amish country of central Pennsylvania (mostly barns) to the modern log home industry spread out across the United States. Buildings

FIGURE 1.1 Buddhist temple Hōryūji. The wood used to build it is 1,300 years old. Wikimedia commons.

made in this traditional style serve as a reminder to architects and designers of the enduring beauty and utility of wood.

But whenever architects consider using wood as the primary construction material, the issue of fire inevitably arises and justifiably so. Fires have consumed whole cities, all of which were made of wood and fell to ash within days (e.g., London, 1666; New York City, 1835; San Francisco, 1849–1851 and again in 1906; Chicago, 1871). More recently, fires that swept through central and northern California consumed a large part of the city of Santa Rosa (see fig. 1.2) and all of Paradise.

When fire breaks out in a mature forest, the outcome is notably different. Tree trunks do not typically succumb to fire by completely combusting. Instead, their outer bark burns off, leaving the

FIGURE 1.2 A devastated portion of Santa Rosa, California, after a major fire in 2017. Air National Guard photo by Master Sgt. David Loeffler

densely compacted cellulose and lignin heartwood unaltered. Only the surface of the exposed trunk is charred. Observe the aftermath of any forest fire if you doubt the resilience of a tree's core.

One year after a forest fire, the tree trunks are still intact. All of the leaves are gone, and most of the branches are fragile, exposed extensions, but the lifeless trunk remains upright (see fig. 1.3). That is because fire needs more than just fuel (i.e., organic matter); it also requires an oxidant. When fire encounters a dense

FIGURE 1.3 Aftermath of a forest fire. Note that the tree trunks are charred but still remain intact. USDA, photo by Morris Johnson.

portion of wood, it stops dead in its tracks. The burned surface prevents the fire from penetrating further, and the conflagration quite literally runs out of gas (in this case oxygen).

Nature's lesson is clear. If you replicate this tree feature by creating a dense wood product equivalent to its trunk, you can construct a building that will not be destroyed every time fire breaks out, all the while capturing tons of carbon (wood is 50 percent carbon by weight) in its very structure—floors, ceilings, roof, and siding. Enter cross-laminated timber (CLT) as a viable option. A building made of CLT becomes the technological equivalent of a tree's carbon storage potential. That is why I am so passionate about its use in the construction of our future metropolises.

Making a city out of fire-resistant carbon-rich material also makes good ecological sense and is a critical pillar in my manifesto for urban change.[3] The manufacture of concrete and steel has a substantial carbon footprint and is responsible for approximately 8 percent of atmospheric carbon. When cities begin to lock up carbon in the form of engineered wood as an integral part of their infrastructure, not simply in thin, dry, and flammable plywood and two-by-fours, it offers hope that our atmosphere will be able to regain its equilibrium with the land and oceans. It is important to emphasize that in properly managed forests, new trees will grow up to replace those used as the source of CLT and other engineered wood products. By the time these new trees mature and are harvested, in about twenty years, even more CO_2 will have been removed from the atmosphere. In well-managed large forests, the process of regrowth is matched by the rate of harvesting. The building industry finally has at its disposal a renewable resource that takes advantage of the forest's finely honed repair and regenerative mechanisms.

New technology can allow us to stop altering the rate of climate change because of our building processes. That goal might not take longer than a hundred years or so if we act soon. Many architects, planners, and schools of forestry and architecture are also promoting the increased use of wood in the urban landscape to help slow down the rate of climate change.[4] Some construction professionals even refer to CLT as the "new concrete."[5]

A brief look back at the milestones in the development of engineered wood products will illustrate how we got to this point. One of the first such products to see widespread application was plywood, developed in Europe in the early 1800s. Its many advantages in erecting various building types were immediately apparent, and plywood rapidly became a standard construction material. In 1865, the United States imported its first plywood sheets. By the 1920s, the United States had adopted standardized methods for the production of plywood that led to the commercialization of the now-familiar four-by-eight, five-ply product. The global construction industry depends on plywood because it is affordable, versatile, durable, and readily available. One of its significant drawbacks is that it is not resistant to fire because the layers are only a fraction of an inch thick and not nearly dense enough to resist total combustion. What's more, its thin layers are held together with organic adhesives, some of whose components (e.g., formaldehyde) are known carcinogens. When plywood burns, it releases toxic chemicals into the environment. These two strikes against traditional plywood panels rule it out for helping us build our new city.

Recent innovations have resulted in the invention of mass plywood panels (MPP) (see fig. 1.4). By layering three-quarter-inch panels on top of one another and alternating each panel by 90°, a panel can be up to twenty-four inches thick and offer similar advantages to CLT with respect to its ability to resist total combustion during a fire. The switch to food-grade polyurethane-based

FIGURE 1.4 Cross-laminated mass plywood panels. Courtesy Freres Lumber.

adhesives in its manufacture allows us to include this new wood-based building material in the growing arsenal of engineered wood products suitable for use in the construction of buildings in the future metropolis.

The use of whole logs for fashioning the frame of a wide variety of structures continues today. For centuries, entire logs were the common choice of building fabrication. The secret to the durability of these structures lies in the design of the highly versatile joinery, which was first perfected in China. Its resilience and strength rely on ropes, wooden pegs, and cleverly configured interlocking beams, often aided by a complex set of braces, all nested together, integrating the supporting members into the rest of the structure without the use of nails or glue (see fig. 1.5).

FIGURE 1.5 A variety of traditional wood braces. Photo by TANAKA Juuyoh (田中十洋)

Gravity alone is responsible for the joinery and braces tightening up over time. As the building sways back and forth, responding to an ever-changing environment, the strength of the overall structure increases, enabling it to survive for centuries. Some of these exceptionally well-constructed buildings are found throughout China and Japan. A few have even escaped the damaging effects of countless earthquakes, many of which exceeded magnitude 7. In contrast, traditional construction emphasizes rigidity and has the opposite effect. Its failure to absorb and dampen the energy of an earthquake places stress on the entire structure, and with enough force, it could and often does cause them to collapse.

In a documentary first aired on PBS, *Secrets of the Forbidden City*, a 1:5 scale model of one of the main buildings was used in a series of tests to see what it might take to destroy it.[6] No matter how hard the scientists and engineers tried, it withstood the shaking forces of the machine's platform, exceeding the highest magnitude earthquake (9.5) ever recorded for that area of China. When the seismic platform was ramped up even higher—to magnitude 10.1, the upper limit of the instrument's ability to replicate an earthquake—the model held its own and stood in defiance of one of nature's most powerful destructive forces. It is a superb demonstration of the value of overengineering to avoid the worst-case scenario of an anticipated disaster. Many of the Forbidden City's structures were laid waste by fires caused by lightning, but they were rapidly rebuilt. The average thickness of most of their framework and nonsupporting materials was not dense enough to simply char. Buildings made with CLT overcome this drawback, as they are explicitly designed with density as their main feature.

Before the era of wholesale deforestation began, catalyzed by the invention of dynamite in 1866, the abundance of old, large-diameter trees allowed for the continuation of the traditional timber frame industry. Still in use are medium-sized logs that are plentiful throughout most of North America and Europe. Today, large-diameter beams are fashioned out of much smaller ones by using engineering, computer-guided saws, and safe-to-use synthetic adhesives. The result is an extensive menu of wood products that includes CLT, MPP, glued laminated timber (referred to as glulam), and nail-laminated timber (NLT).

Bamboo, a grass, can also be engineered and is gaining in popularity as a material for construction. Many mass bamboo timber (MBT) products are commercially available.[7] Its advantages include that it has a high strength to weight ratio and is lighter and stronger than wood, reducing the weight of load-bearing columns by 20 percent compared to cross-laminated timber. Bamboo is the fastest-growing terrestrial plant on Earth. Some species grow three feet per day, making MBT one of the most desirable renewable building materials so far developed.[8] It is also useful in the phytoremediation of damaged forest ecosystems. However, a significant disadvantage of using MBT is its susceptibility to a wide variety of fungi and insect pests. Another drawback is bamboo's tendency to spread out through its root systems and invade adjacent patches of undisturbed forest.

ENGINEERED WOOD PRODUCTS

Cross-Laminated Timber

Cross-laminated timber (see fig. 1.6) was invented in 1994 by Gerhard Schickhofer, then a graduate student at the Graz University of Technology in Vienna, Austria.[9] Its first application to the built environment occurred in 2002, and it slowly became accepted in Europe as a viable substitute for structures originally designed to be made from concrete and steel. Today, CLT is certified for use in all construction projects by most municipal building codes throughout Europe. It is also becoming widely used in Canada and the United States.[10] In 2015, a revision of the U.S. National Design Specifications building codes approved engineered wood as a mainstream building material. Convincing data for that

FIGURE 1.6 Cross-laminated timber panels. Courtesy USDA

change were generated from a series of construction industry facilities that put CLT through their fire resistance protocols and showed that it uniformly exceeded the limits of the building code standards. The use of CLT dramatically increased shortly after it was approved in 2018 as a standard building material by the International Code Council:

SECTION 602.4 TYPE IV (*MASS TIMBER*)

Type IV construction is that type of construction in which the building elements are mass timber or noncombustible materials and have fire-resistant rating in accordance with Table 601.

Mass timber elements shall meet the fire-resistance rating requirements of this section based on either the fire-resistance rating of the noncombustible protection, the mass timber, or a combination of both and shall be determined in accordance with section 703.2. . . .

In buildings of Type IV-A, IV-B, and IV-C construction with an occupied floor located more than 75 feet (22,860 mm) above the lowest level of fire department access, up to and including 12 stories or 180 feet (54,864 mm) above grade plane, mass timber interior exit and elevator hoistway enclosures shall be protected in accordance with Section 602.4.1.2 in buildings greater than 12 stories or 180 feet (54,864 mm) above grade plane, interior exit and elevator hoistway enclosures shall be constructed of noncombustible materials.

CLT as a construction material has many advantages, the most important one being that it is a renewable resource. The mantra of the CLT industry is *Lighter, Stronger, Cheaper, Smarter, Faster, Healthier.* Typically, CLT is precision-manufactured off-site as prefabricated units. The various components are then delivered to the building site and installed, usually within days of being produced. This dramatically reduces the time and cost of construction and eliminates the solid and liquid wastes created by conventional on-site construction activity. Usually, buildings made of CLT take weeks to complete, compared to months or even years with large concrete and steel structures. CLT is, pound for pound, twice as strong as concrete or steel and, by volume, only 20 percent as heavy as reinforced concrete, making it safe and easy to manipulate during the construction process.

Because CLT embraces the centuries-old design of timber-frame joinery, erecting a CLT building is reminiscent of building with a set of Lincoln Logs. In today's joinery schemes, the addition of high-strength, lightweight metal braces that secure the weight-bearing parts of the structure makes the building even more resilient. The ease of recycling CLT for reuse is another of its many advantages, making it far superior to concrete and steel. In fact, concrete can be recycled only as aggregate or in landfills, and steel girders usually must be melted down for reuse, consuming a large quantity of energy in the process. Several other desirable features of CLT include its sound-dampening qualities and its response to seasonal changes in humidity, which enable it to cool interiors in summer and retain heat in winter.

Buildings made from CLT store carbon, and they are also aesthetically pleasing. Entering a modern all-wood building, one encounters an environment warm in tone and nurturing to the senses in a way that is unrivaled by any other building material (see figs. 1.7, 1.8). Most would agree that an occasional walk in the woods or tree-filled park is stimulating, refreshing, and soothing. Imagine if those same feelings welled up every time we went to school, to work, or to some inviting, innovatively designed restaurant built with CLT. It is easy to undervalue the importance of artistic sensibilities when a builder or developer considers a project. Still, with CLT, one can attain high marks even from those architects and builders who devote most of their attention to finances. Lastly, the cost of producing CLT panels is lower than an equivalent amount of concrete. This fact alone should go a long way to convince the construction industry.

Making CLT begins by harvesting trees, preferably from forests managed as much for their biodiversity as for their production of wood. These are not mutually exclusive goals: this practice is now firmly established and successful worldwide. In one study, forests managed for biodiversity and lumber production increased the number of harvestable trees.[11] Any species of pine around nine inches in diameter is the preferred starting material. In the United States, Douglas fir and spruce are the trees of choice.

Larger trees are typically left standing, which is fortunate since their presence is essential to maintaining the forest's ecosystem.[12] They are the institutional memory of the woodlands, and Suzanne Simard (see introduction) refers to them affectionately

FIGURE 1.7 Interior of the Ascent building. Kevin Marshall, photographer.

FIGURE 1.8 Interior of a home made from cross-laminated timber. Courtesy Diana Snape, architect

as "mother trees." Keeping them alive ensures the continued reseeding and good health of the forest. No planting is necessary.

Cutters strip timbered trees of branches and bark and then mill them into 2 × 5 or 2 × 6 inch boards. They are then run through a planer and made square. After kiln drying, the boards are ready for joining into panels. If longer boards are required, finger joints are cut into the ends of shorter ones. A rapidly curing adhesive is applied to both ends of the finger joints, and the boards are fitted together. When the glue has set, the longer boards are cut to specified lengths.

Many milled boards of similar size are placed in a row to form a panel after their two-inch-wide edges are coated with an adhesive. The panel is subjected to pressure from both sides, ensuring a tight bond between each board. Individual panels are then face-glued to each other so that the grains of one are oriented 90° to the next one (hence the term *cross-lamination*). CLTs of three, five, or seven panels are formed, creating a durable, lightweight, fire-resistant, carbon-storing building material. After trimming each panel to an approximate size, they are shaped according to a computer numeric control (CNC) machine, a sophisticated computer-guided saw. Soft, muted patterned interior designs can be achieved by taking advantage of the grain and other tree features inherent in the CLT panel's wood structure.

Nail-Laminated Timber

Nail-laminated timber preceded the invention of CLT by nearly a hundred years. In contrast to the manufacture of CLT, NLT requires no gluing. Instead, nails connect segments of wood (usually 2 × 4 inch boards) to form a single panel (see fig. 1.9). NLT panels are not suited for use as weight-bearing members.

FIGURE 1.9 Nail-laminated timber panel. Courtesy Think Wood

According to Think Wood, an industry advocacy group for engineered wood products,

> To create NLT, dimension lumber is placed on edge with individual laminations mechanically fastened together with nails or screws. The boards are nominal 2×, 3×, and 4× thickness. Width is typically 4–12 inches. NLT gets its strength and durability from the nails/screws that fasten individual pieces of dimensional lumber into a single structural element.
>
> NLT's revival is due in large part to domestic availability. Compared with other building materials like CLT, this mass timber product does not require a dedicated manufacturing facility—and it can be fabricated with readily available dimensional lumber. This allows project teams and manufacturers to use locally sourced materials.
>
> Applications for NLT include flooring, decking, roofing, and walls, as well as elevator and stair shafts. Because NLT is made of wood, it offers a consistent and attractive appearance for decorative or exposed-to-view applications. The International Building Code (IBC) recognizes NLT as code-compliant for buildings with varying heights, areas, and occupancies, allowing for Type III, Type IV, or Type V construction.
>
> Architects like NLT and use it to create monolithic slab panels that support various structural and design needs, including curves and cantilevers. The addition of plywood or oriented strand board sheathing on one face of the panel provides in-plane shear capacity, allowing NLT to be used as a shear wall or structural diaphragm.

The first tall structure in America using NLT for the interiors was the T3 building (Timber, Transit, Technology) erected in 2016 in Minneapolis and designed by Michael Green (see fig. 1.10).[13] The seven-story building is clad in weathering steel. Remarkably, the T3 building weighs one-fifth as much as an equivalent structure made from concrete and steel beams and stores 3,646 tons

FIGURE 1.10 T3 building, Minneapolis, Minnesota. Michael Green, architect; photographer, Ema Peter

of carbon in its structure. While NLT is still in use, other more versatile iterations of engineered wood, particularly CLT, have taken some of the luster off NLT as a building material.

One interesting aspect of the wide versatility of engineered-wood construction is that certain damaged trees can be used without compromising any of the physical aspects of the wood. Michael Green harvested trees in the Pacific Northwest region that were killed by the mountain pine beetle (see fig. 0.12) and used them for a significant portion of the T3 building.[14]

FIGURE 1.11 Model-C building, Roxbury, Massachusetts. John Klein, architect.

Most architects and builders who have worked with engineered wood are as enthusiastic about its role in carbon storage as they are about its ease of use. Among the numerous structures made of CLT—the Model-C building (see fig. 1.11), the Forte building (see fig. 1.12), the STEM building at Michigan State University (see fig. 1.13), and Milwaukee's Ascent (see fig. 1.14), the tallest CLT building in North America—serve as outstanding examples of the use of engineered wood products.

Daltston Works, an apartment complex in London (see fig. 1.15) built in 2017, is a stylish, modern expression of integrated living spaces, setting the tone for the future of urban living. Its structure captures 2,600 tons of carbon. A BBC interview from 2019 features the architect, Andrew Waugh:

The average lifetime of a building is 50–60 years and is more than enough time for architects and engineers to work out the

FIGURE 1.12 Forte building, Melbourne, Australia. Lendlease Architects.

FIGURE 1.13 Michigan State University S.T.E.M. building, East Lansing, Michigan. Integrated Design Solutions and Ellenzweig Architects.

re-use and recycling issues (of CLT). Turning it into biochar is one possibility." Waugh's buildings are made to be easy to take apart for re-use by future generations. Fundamentally he—along with a growing group of international architects is convinced that mass adoption of CLT is an important weapon in the fight against climate change. "It's not a fad or a fashion," he tells me as we finish the tour of his East London build, and I take my final, incongruous breath of the forest air. "The largest commercial developer in the UK has just bought this building. For me, that's where you want to be . . . I want this to be main-stream. Everybody should be building with this." I return to my original question: could we realistically return to wood as our primary building material? "It's not only realistic, it's imper-ative," argues Waugh. "It has to happen. In architecture you always go back to the sketch: the sketch is climate change.[15]

FIGURE 1.14 Ascent building, Milwaukee, Wisconsin. Korb + Associates Architects. Chris Lark, photographer.

FIGURE 1.15 Daltston Works apartment complex constructed out of cross-laminated timber. Courtesy Andrew Waugh, architect.

Several new engineered wood products have become commercially available. Dowel-laminated timber (DLT) is made from 2 × 4-inch cuts of wood. Each board is placed on edge and then a row of holes is drilled. All boards with holes are joined together by friction, using a long, slightly wider hardwood dowel. The dowels are driven through all the holes in each board to form a single panel. DLT is faster and cheaper to manufacture, and since no glue is needed, after assembly the timber is immediately ready for use. DLT is used mainly for floors and ceilings. Laminated veneer lumber (LVL) panels are made just like CLT panels, and in addition a thin layer of hardwood (e.g., maple, walnut, or ash) is glued as the top layer. LVL is used exclusively for interior work. Finally, parallel-strand lumber (PSL) is made of bundles of long thin strands of fir, western hemlock, or pine. The strands are glued together in parallel under high pressure. PSL is ideal for use as beams and columns. The use of bundles of wood strands is analogous to the use of strands of metal wire employed in building suspension bridges. When many strands are woven together, they create a strong structure.

The weight advantage of engineered wood products over conventional construction materials continues to boost the materials' popularity as more varieties become available. They even have the potential to significantly expand the range of places where the construction of large buildings is possible. For example, bedrock dominates the geology of Manhattan island, and it's mostly found at or very near the surface. But in the midtown area, the bedrock is deeply buried under sand and limestone deposits that cannot support tall buildings made with traditional building materials.[16] CLT buildings could represent a potential solution to this problem.

New York is one of the best examples of a modern metropolis. Its area is just over 302 square miles, and it comprises five separate boroughs: Queens, the Bronx, Manhattan, Brooklyn, and Staten Island. Some eight million people live there and take advantage of the many attractions and services a large, diverse city has to offer. It has one of the world's first subway systems, and for the most part, it operates efficiently, enabling all New Yorkers a convenient ride to all parts of the city except Staten Island, which is easily reached by ferry. Its many parks, zoos, and public spaces attract millions of visitors, and it has long been a world destination for tourism. Founded in 1624 by the Dutch and named New Amsterdam, the British took it over in 1664 and renamed it. Two of its oldest remaining buildings are the Wyckoff House in Brooklyn, built in 1652, and Fraunces Tavern in Manhattan, originally built as a home in 1716.

If we were to travel back in time to New York in the year 1900 and then return to the present, we would see that fewer than 20 percent of all the buildings present in 1900 still exist. What remains is a small, odd collection of old buildings, reminders of the city's storied past. This one observation is enough to allow an estimate of 100–150 years to convert New York into a city with most of its structures made of engineered wood.

An in-depth meta-analysis has been carried out by Ali Amini and his colleagues on the atmospheric carbon sequestration potential of replacing old buildings with those constructed using engineered wood over a twenty-year period. They conclude:

> To ensure a reliable estimation, 50 different case buildings were selected and reviewed. The carbon storage per m² of each case building was calculated and three types of wooden buildings were identified based on their carbon storage capacity. Finally, four European construction scenarios were generated based on the percentage of buildings constructed from wood and the type of wooden buildings. . . . The annual amount of CO_2 captured would then be 2 Mt for 2020, 15 Mt for 2030, and 55 Mt for 2040. Thus, the cumulative amount for this 20-year period would be 0.42 Gt.[17]

Expanding this kind of analysis to include all major cities could serve as the impetus for moving toward a future of cityscapes dominated by mass timber buildings.[18]

2

Pillar Two
Urban Agriculture

An area less than one-fourth that which, in my boyhood days, supplied the "garden truck" for the family, will produce foodstuffs of variety, quality, quantity and value never dreamed of by the home gardener.

—WILLIAM GERICKE

A BRIEF HISTORY OF FARMING

No human activity over the past ten thousand years has transformed our planet more than farming. We have cleared away nearly half of the world's forests to make room for food crops and repurposed roughly 3.3 million square kilometers of native grasslands to produce various grains and raise a wide variety of grazing animals (see fig. 2.1).[1]

At the beginning of the agricultural revolution, the amount of farmland was minuscule, but as our population increased, we destroyed whole ecosystems to create more land for farming. Until that point, the plants, animals, and microbial networks had taken millions of years to perfect their biodiversity, ensuring their long-term survival. Then we changed the wilderness into something that served our needs only, monocultures purposely designed to discourage biodiversity. Our alteration of the land to accommodate all forms of agriculture created adverse, long-term, global consequences for our climate.

Farming first arose in modern Turkey, Jordan, Iraq, and Palestine. From there, it rapidly spread to Egypt and other parts

61

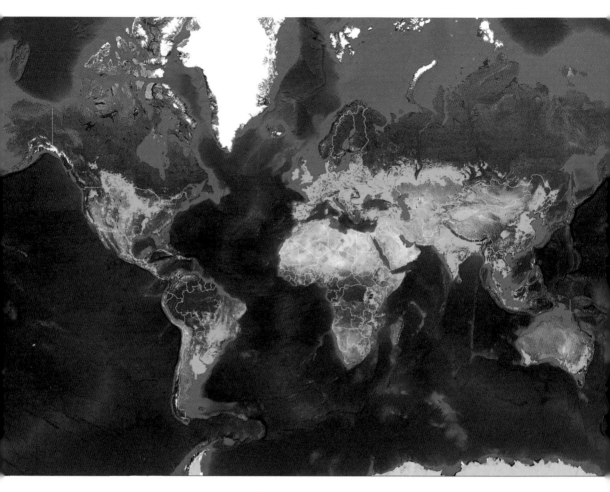

FIGURE 2.1 Global distribution of agricultural land. Courtesy United States Geological Survey.

of the Middle East. Some of the most productive farmland lay along the floodplains of rivers. The Nile, Tigris, and Euphrates were pivotal in establishing highly productive farming practices. Nutrient-laden silt was deposited on their banks when they overflowed during the rainy season. When agriculture first took hold in Egypt and Mesopotamia, annual flooding was a regular occurrence only in Egypt, and the early farmers were wholly dependent upon it for their crops. As agriculture flourished, so did the

Egyptian civilization. When there was an occasional drought, crops failed, and populations suffered. Irrigation systems were not invented until around 6,000 BP.[2]

The history of those two iconic regions played out differently because one lacked a sustainable food supply. The early farmers of Mesopotamia's Fertile Crescent were forced to keep moving to new land further south. Wheat crops along the two rivers' broad floodplains rapidly exhausted the nutrient-rich silt, whose replenishment was more erratic than along the banks of the Nile. The unpredictable annual cycle of precipitation, resulting in massive, prolonged loss of crops, doomed entire cultures to extinction.[3]

Earth's changing climate is the norm, but early farmers had no experience in dealing with an unpredictable environment. Relying on what they presumed were "dependable" annual cycles of precipitation for growing their crops left them at the mercy of natural processes.

The first crops led to the establishment of the first cities, which sprang up next to the land that produced them. For example, einkorn wheat (Triticum monoccocum), a wild grass, was domesticated by farmers in the Middle East eleven thousand years ago. Through trial and error, farmers eventually learned how to breed plants for characteristics that gradually improved the size and the number of grains per stalk. Jericho, founded at about the same time, became the world's first metropolis. Abundant springs supplied the residents with a steady flow of fresh water. Jericho is located near the west bank of the Jordan River, an essential ingredient in establishing the world's first irrigation system.

People soon realized that grains of all kinds were more than just a source of nutrition because they could be stored for years if necessary. Not long after the invention of agriculture, wheat, barley, and rice came into use as money. Trade routes developed in many places adjacent to these early urban and agricultural centers, and so began the start of modern civilization.

The ruins of many ancient cities, however, show that farming was not always successful. Ur, once the capital of the Sumer civilization, was once the world's most populated urban center, home to fifty thousand people. Then a three-hundred-year-long drought affected the course of the Euphrates, stranding Ur and a significant number of other founding cities without an easily accessible source of fresh water. Climate change made cities routinely fail as the farms that fueled their progress encountered long-term adverse growing conditions. Lack of sustainable agriculture led to civil unrest and war. Cities fortunate enough to be able to produce crops were eventually destroyed by plundering armies.

A number of failed cultures (e.g., Mayan, Incan, Khmer, Anasazi) used highly creative irrigation schemes to prolong the productivity of their farmland. But over time, none of those agricultural systems could meet the increased nutritional demands of an expanding population and a changing climate.[4]

During the ensuing millennia, as farming expanded its footprint to accommodate population growth, so did cities. Urban expansion encouraged more commerce, competing with local food production. Farming lost out and was relegated to the surrounding countryside. Today, large-scale industrial agriculture is typically located hundreds to thousands of miles away from cities with populations exceeding half a million, and getting food to the table has become a logistical nightmare.

As of 2022, eight billion people relied on their food supply coming from farmland that totaled over five billion hectares (twenty million square miles) (see fig. 2.1).[5] But farming at that scale inevitably creates environmental problems. In 2018, agriculture was responsible for 5.8 billion tons of methane and nitrous oxide (i.e., non-CO_2 greenhouse gases or GHGs).[6] Both are major contributors to climate change. Activities that produced them included rice farming, cattle husbandry, and artificial fertilizer production.

TRADITIONAL FARMING TODAY

The single most important confounding factor in predicting global food production is climate change. But an average estimate of yields for a wide variety of essential crops is still possible if the data are available. Under the best circumstances, when precipitation and temperature profiles over the growing season are normal, most commercial crops yield between 50 and 70 percent of the total seeds planted. But that is when agrochemicals (pesticides and herbicides) are employed to manage insect pests and weeds. For small stakeholder farms, the kind found in most parts of the less and least developed world, yields average just under 40 percent even in good years.

Globally, crop losses now occur regularly due to the increased frequency of unfavorable weather events and a more volatile climate. Reduction in yield of pre- and post-harvested crops, especially grains, due to a wide spectrum of unwanted diners—vermin, insects, and microbial plant pathogens—also contributes to crop losses, negatively impacting long-distance food supply chains. Loss of harvest is not only measured by a drop in yield; storage and shipping both reduce the nutritional value of some crops, such as leafy green vegetables.[7]

Unless we can dramatically reduce losses due to post-harvest spoilage and waste after consumers purchase their food, production will need to almost double in twenty years to keep pace with the increasing population. Before that happens, we will reach a tipping point caused by the lack of available new farmland.

Deforestation is one of the main human activities responsible for RCC. Nearly three trillion trees are missing from our planet because of our collective agricultural endeavor. There are approximately three trillion trees still standing, so we can surmise that the Earth's ability to sequester carbon in the form of trees has been reduced by about 46 percent. Is it any wonder that the atmospheric content of CO_2 (one of the leading greenhouse gases responsible for atmospheric warming) rose nearly 50 percent from historically low levels of roughly 280 parts per million (from the level ten thousand years ago up to the advent of the Industrial Revolution) to 415 parts per million (see fig. 2.2) in January 2021? Our consumption of fossil fuels caused that increase.[8]

When I first began to focus on environmental issues, I imagined how farming could be carried out without further disrupting terrestrial ecosystems. Forests are still being sacrificed to make room for more traditional farms, yet trees are our best hope for capturing excess atmospheric carbon dioxide. Vertical farming, urban agriculture carried out in tall buildings without soil, is one solution. It can potentially help cities join the global effort to help slow the rate of climate change by allowing forests to recover and perhaps even expand back to their original footprint.

Suppose indoor urban farms that take up less land were to replace a large portion of traditional agriculture. Fragmented ecosystems could eventually repair themselves and resume their role in creating habitats for millions of life forms. When we destroyed those habitats by repurposing land for farming, nature changed direction, and the climate began to push back, telling us to stop, but we did not pay attention.

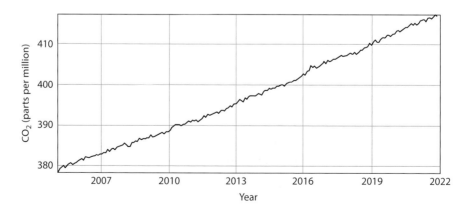

FIGURE 2.2 Carbon dioxide levels in the atmosphere. Courtesy NASA.

Abandoned farms are common in many parts of the world. Their failure to survive is due to many things, among which are changes in weather patterns. Securing more land by deforestation to compensate for those abandoned farms only exacerbates an already out-of-control situation.

I have spoken at eco-summits about alternative farming strategies, with an emphasis on vertical farming. While the industry has made progress in creating indoor growing systems with high yields (mostly leafy greens), it is still in its earliest stages of development. I would like to see controlled environment agriculture (CEA) make significant inroads in replacing most traditional outdoor farming, especially in regions where annual crop failures are the norm. But we cannot rely solely on CEA for the solution to the carbon storage problem. Chapter 1 addresses this issue with a potential solution inspired by nature; emulate temperate zone forests when we replace old builds in the urban environment with new ones.

FOOD AWARENESS AND URBAN AGRICULTURE

Urban populations are becoming more concerned about where their food comes from and what's in it. The rate of outbreaks of food-borne infectious diseases and the health risks from consuming contaminants on and inside fruits and vegetables grown in soils polluted with agricultural and industrial chemicals have significantly increased over the last ten years. Food safety awareness has spread worldwide. In reaction to global food insecurity issues, a proactive segment of the public has begun to advocate for changes in the way food is produced and how it is monitored for the presence of noxious agents.[9] Healthier, safer, and fresher food is their call to arms. It has even motivated some to become city farmers despite any real support or encouragement from most municipal or federal governments. Japan, Singapore, and the Netherlands are exceptions, encouraging the development of all forms of indoor farming.

Planting crops in abandoned lots, rooftop gardens, greenhouses on top of supermarkets, restaurants with indoor farms, and large industrial vertical farms grew out of this movement, driving the exponential growth of urban farming innovations. Since 2010, the number of vertical farms has increased dramatically, and examples exist in most countries of Europe, North America, the United Arab Emirates, India, Taiwan, and China (see fig. 2.3).[10] Market research estimates project that by 2030, the collective value of the industry will exceed $31.15 billion.[11]

Replacing a substantial number of traditional farms with other forms of urban agriculture in every city would allow for the reforesting of damaged ecosystems. No technological barriers prevent

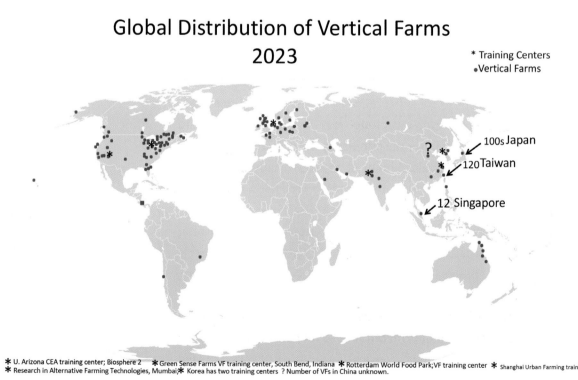

FIGURE 2.3 Global distribution of vertical farms in 2022.

us from implementing this plan. The goal is to restore the ecological balance between the natural world and ours. For our planet to foster the repair, proliferation, and maintenance of wild places, a radical change in the way we interact with nature is required.

Hyperurbanization has overwhelmed many cities in some parts of Asia. Hundreds of thousands of displaced farmers now live in cities. The option for continuing their livelihood in an urban setting by becoming city farmers may lead to the widespread adoption of farming, in a much safer and more efficient way, indoors.

VERTICAL FARMING

Vertical farming enables cities to repurpose obsolete buildings into centers for urban agriculture. AeroFarms in Newark, New Jersey, is an excellent example of this kind of real estate rebirth. An abandoned ironworks complex was transformed into a highly productive indoor farming facility that now employs over 150 residents. Property values in the surrounding area go up the moment a vertical farm succeeds. Mayors of cities looking to increase their corporate tax base are encouraged by the growth of urban agriculture. They frequently offer long-term financial incentives, and many are willing to lower the purchase price of an empty building to encourage the establishment of this novel industry. Vertical farming creates new jobs while providing a reliable and safe food supply that operates year-round.

An important difference between outdoor and indoor farming is in a crop's percentage yield. Typically, hydroponics and aeroponics are similar in this respect, with 85 to 95 percent of the seedlings maturing to harvest. Being able to increase production without increasing the physical footprint of the farm—by growing up instead of horizontally—is an enormous advantage over traditional farming. Vertical farming can be established virtually anywhere in the world provided sufficient freshwater is available.

Methods for producing indoor crops have become more efficient and easier to implement over the last ten years, creating
the foundation for a robust, diverse supply chain for customized
grow systems, lighting modalities, nutrient solutions, AI-driven
surveillance (quality control) devices, and harvesting and pack-
aging strategies. Many annual trade shows featuring indoor
farming industries have registered an increase in manufacturers
of each of these CEA segments, and all have been well attended,
with the exception of the recent pandemic years.

Designing the majority of the new city's buildings with some
form of vertical farming capability will set an example for other
metropolitan centers. The city will mandate and help facilitate
the inclusion of some form of indoor agriculture in restaurants,
apartment complexes, and office buildings (see figs. 2.4a, 2.4b).

FIGURE 2.4A Pasona O2 building, Tokyo, Japan.

FIGURE 2.4B Indoor rice field in lobby of Pasona O2 building.

Seasonal menus offered by many upscale restaurants only contain food items available from local farms. Soon, these farms will have to compete with those that produce a wide range of crops year-round in CEA facilities. There are many excellent examples of each prototype building with built-in food-growing capacity already up and running.

The varieties of crops produced will depend largely on the cuisine preferences of a given city's population. Virtually any plant can be grown indoors. Food can be as fresh as one hour old from harvest to consumption. The full nutritional value of each crop will be realized. All major edible plant varieties will be available year-round, making dining in as rewarding as dining out. Exchanges of crops between buildings, neighborhoods, and districts will enhance the experience of urban life, creating future generations of healthier, food-savvy diners. The produce markets in our new city will become the playground for a new breed of chefs who will avail themselves of any foods that can be grown and incorporate them into their culinary repertoire. The public will be treated to a dazzling array of tasty, innovative, healthy menu choices.

Designing buildings that accommodate food production and all the other characteristics required to get them off the municipal grids (energy and water) may seem challenging. Indoor farms must be able to prevent uninvited visits from curious bystanders and insect pests that want to sample the edibles. Fortunately, an effective barrier system is not difficult to achieve. Numerous vertical farms have incorporated crop security systems into their facilities; only specific personnel are allowed to access the grow rooms, and the air inside is lightly pressurized to keep out plant pathogens. In addition, creating walk-through, secure double entrances to the growing rooms and mandating that staff wear hairnets, shoe coverings, gowns, and gloves constitute a proven protocol for maintaining crop safety.

Hydroponics (see fig. 2.5) uses 70 percent less water than outdoor farming for the same crop, and aeroponics (see fig. 2.6) uses 70–90 percent less than hydroponics. Grow racks can be customized for any building, and many lighting schemes have been engineered so they can be fine-tuned to specific wavelengths by adjusting modular LED wavelength bands and voltage-controlled irradiance compatible with the physiological requirements of individual crops.[12] Lighting research continues to yield useful data regarding the optimization of indoor LEDs for specific crops.

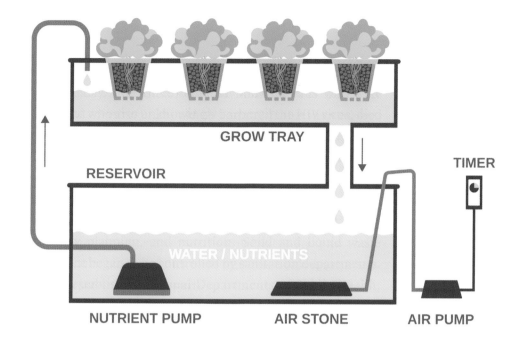

FIGURE 2.5 Schematic diagram of a hydroponics system.

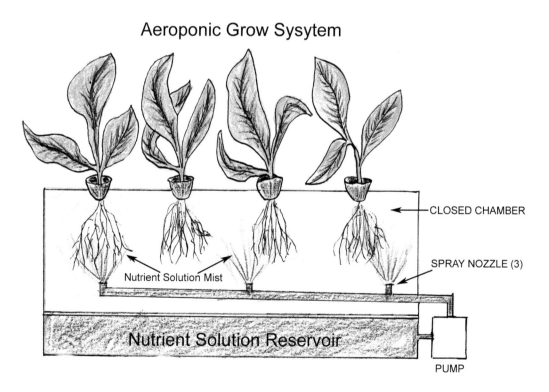

FIGURE 2.6 Schematic diagram of an aeroponics system.

The resources needed to accommodate CEA include freshwater, nutrient solutions, electricity, and HVAC systems that keep the ambient temperature between 60 and 85 °F. The amount of each resource will depend on the size of the farm and its location.

Urban agriculture, when widely applied, will significantly reduce a city's dependency on the surrounding landscape. Each acre of indoor farming will yield the equivalent of many acres of outdoor farming. The actual number of indoor farms required will depend on the kind of crop and the method employed for its production. It will take an enormous number of vertical farms to replace substantial amounts of farmland, which seems like a daunting task, but over the next fifty years, it is predicted that climate change will cause an ever-increasing number of crop failures due to the unpredictability of annual precipitation in areas where agriculture is now succeeding. In turn, the loss of traditional farmland will accelerate the scale-up of indoor farming to the built environment.

Abandoned farmland typically reverts back to its original ecological function.[13] Depending on the location, this can mean more trees. The five billion hectares of agricultural land currently needed to feed the world could be reduced by nearly half if every city produced a substantial portion of its plant-based food supply. It would then be possible for nature to regenerate the world's damaged and fragmented forests. The regrowth of old forests decimated by the aggressive, short-sighted acquisition of farmland that was later abandoned has been documented many times in the past fifty years. It is one of nature's primary ways of expressing resilience. To encourage this process, we simply need to leave the land alone. A change in policy at the international governmental level might be required to establish a global land trust. It would serve a dual purpose: reforestation for carbon capture and creation of a renewable resource for selectively harvesting trees, the raw material for the manufacture of engineered wood products.

Urban agriculture holds the promise of being able to create an ample, varied, safe food supply, while at the same time enabling the rewilding of abandoned farmland. Encouraging farmers, who often struggle to make ends meet, to become carbon farmers by growing trees, and paying them a fair wage for that effort, will allow them to remain on their land and continue to take responsibility for its stewardship.

Trees produce food both for themselves and for a wide variety of wildlife. Nuts, fruits, and even leaves provide ample nutrition for an astonishing variety of invertebrates and both cold- and warm-blooded vertebrates. Urban agriculture reflects this essential function of trees by providing a new local source of food for the new city.

This plan should have a good chance to succeed by expanding our vision of how we want the world to look, not just for us, but for future generations; then we just need to work hard to make it happen.

3

Pillar Three
Harvesting Water from the Air

Till taught by pain,
men know not the value of water.

—LORD BYRON

HISTORY OF WATER USE AND THE RISE OF PUBLIC HEALTH

Water is one of the essential ingredients enabling life on Earth to exist. Nearly all ancient cities, especially those in the Middle East, ultimately were unsustainable because, in the long term, they could not supply enough freshwater to meet the needs of their burgeoning populations. A changing climate, spread out over several hundred years, was the main reason why many of those cities, whose names are now largely forgotten, were abandoned in favor of greener pastures.[1] The establishment of a reliable water supply and the creation of the first sanitation system would have to wait for the rise of the Roman Empire, at least for Western cultures.

By the time Rome reached its zenith around 117 AD, it had built eleven aqueducts constructed out of concrete and stone (see fig. 3.1) to bring drinking water from the mountains to the city's center.[2] Remarkably, many of those aqueducts are still functional. Having access to clean drinking water was a major achievement; the frequency of outbreaks of dysentery and other life-threatening water-borne illnesses diminished dramatically, reduced

77

FIGURE 3.1 Roman aqueduct, constructed circa AD 117. Library of Congress.

to sporadic events but not entirely gone.[3] Everyone in Rome could access the water from the numerous outdoor public fountains strategically located throughout the city. Trevi Fountain (see fig. 3.2), one of the more poetic examples, is still in operation.

Like the aqueduct system, the Roman baths (see fig. 3.3) were highly popular and much appreciated. Regrettably, their continuous use turned them into what one historian described as cesspools.[4] The sewers that ran under the streets (the *Cloaca Maxima*) appeared to be well constructed. But heavy rains easily overwhelmed them, and the pipes often burst at the seams, the raw sewage filling Rome with the stench of pollution. Those spills rapidly became foci for water-borne infections.[5]

Infrastructure maintenance was problematic, as the city seemed to initiate construction projects each time a new emperor

FIGURE 3.2   Trevi Fountain, Rome, Italy.

was crowned. Most residents knew that there was an association between some illnesses and tainted water, but the true causes remained a dark secret. The germ theory of disease would not be proven until the late 1800s. What is more, many of the pipes that crisscrossed under the cobblestone thoroughfares of Rome were fashioned out of one of the most malleable of all metals, lead. This included those pipes carrying drinking water to the palatial abodes of the most privileged of Rome's elite ruling class.

As Rome's imperial influence spread out over most of Europe and into Asia, the means to secure potable water became a familiar feature of many of its conquered cities. Food systems and health care were also creating (albeit ponderously slowly) a healthier empire. Theoretically, an enlightened public could expect not to die from something they drank or ate. On paper,

FIGURE 3.3 Roman baths (restored), Bath, England. Courtesy Diego Delso, License CC-BY-SA

Rome's engineering ideas were brilliant. But functionally, while the aqueducts worked like a charm, the entire sewage system failed miserably, partly because those in charge of sanitation could not anticipate the many ways infectious diseases were transmitted.

A case in point is malaria, which in Latin means "bad air."[6] In the ancient world, this dreaded disease was commonly believed to emanate from the fog (miasma) that periodically shrouded Rome and other cities throughout the Italian peninsula. Emperor Claudius was committed to improving the port city of Ostia, which lay at the mouth of the Tiber River thirty miles east of Rome. He ordered a nearby swamp to be drained and two breakwaters to be constructed in the adjacent city of Portus. Both schemes proved effective, allowing Ostia to develop into a significant trade destination. Miraculously, malaria disappeared from its immediate environment. The Romans would refer to this event as proof that bad air caused malaria. We know now that draining the swamp removed the breeding grounds for the mosquitoes that carried the deadly malaria parasite. However, the real reason for malarial outbreaks wasn't revealed until 1896. Amico Bignami and colleagues discovered that Anopheles mosquitoes are the vectors for *Plasmodium falciparum* (see fig. 3.4), which causes the deadliest kind of malaria. Ironically, they carried out their research at the University of Rome.

Not knowing how diseases spread throughout a community or what caused them meant that infectious organisms would not just go away. Mitchell lists a few of the parasitic infections that forever plagued Rome as examples of the adverse conditions under which everyone suffered.[7] Many of these pathogens were ingested by eating freshly harvested raw vegetables, indicating that the food Romans consumed was highly contaminated with the infectious stages of geohelminth and protozoan pathogens. It turns out that collectors of human excrement were busy each morning making the rounds of houses and loading up their "honey pots" with the unwanted by-product of yesterday's meals. Then they went to the countryside to sell their fertilizer to farmers; it was an effective agricultural system but seriously flawed by the lack of modern sanitation and public health awareness Today, half of the world still uses untreated human feces and urine as fertilizer, with the same mixed results—good yields of a wide

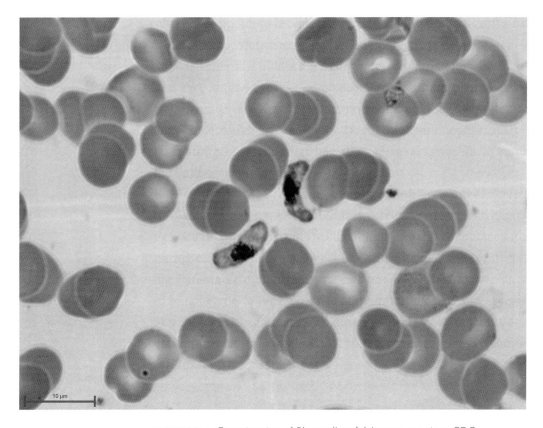

FIGURE 3.4 Gametocytes of *Plasmodium falciparum.* courtesy CDC.

variety of fruits and vegetables and widespread patterns of diseases resulting from a passel of nasty intestinal parasites.[8]

No cities had a solid waste management plan in place. Heaps
of garbage in the streets of nearly all the cities throughout the
Roman Empire was a regular sight. Leftovers were simply
hurled out the window, and a host of ad hoc diners and perid-
omestic animals—cats, dogs, and rats—undoubtedly benefitted
from residents' lack of concern for what to do with their trash.
Rome's rodent population would ultimately play a central role in
the spread of the bubonic plague. At the same time, the Tiber
River served as a convenient liquid waste disposal system. Living

downstream from it was hazardous to one's health, since many villages used water from the Tiber for all their daily needs, including drinking. For a more detailed review of Rome's struggles with sanitation, see Havliček and Morcinek.[9]

If this enlightened metropolis could not prevent the spread of diseases because of its poorly functioning sanitation system, then what would the rest of the world's less developed cities have been like to live in? In a few cases, luck played a role. For example, currents of clean air swept into most coastal cities from across the ocean, which mostly kept aerosols to a minimum. The ocean's tides and currents helped to remove municipal waste. But for the great majority of Rome's residents, living in close proximity resulted in morbidity and mortality rates that limited the average life span to about fifty years.

Overall, right up to the nineteenth century, cities were rife with conditions that challenged a person's ability to live a healthy, normal life. Nonetheless, the lure of employment opportunities, housing, a reliable food source, and multiple entertainment venues enticed most urbanites to stay put.

The introduction of cholera into London's drinking water in 1832 proved to be the seminal event that initiated the development of technologies leading to modern sanitary practices. How this dreaded, often lethal diarrheal disease-causing bacterium traveled from the Bay of Bengal in India to London, England, is a fascinating tale of bad luck, biology, politics, commerce, and clever sleuthing by a curious London physician. For details of the cholera outbreaks in England, see Rosen and Begum.[10]

This is the essence of the story: Cholera (*Vibrio cholerae*) is a microbe (see fig. 3.5) that grows best in brackish water or estuaries (where rivers meet the ocean). Its biological role is to help certain species of zooplankton carry out their reproductive cycles. This occurs in late summer after the estuaries become brackish and laden with nutrients following the rainy season.[11]

Trade between India and England was brokered by the East India Trading Company. In 1831, one of its ships, the *Elizabeth*,

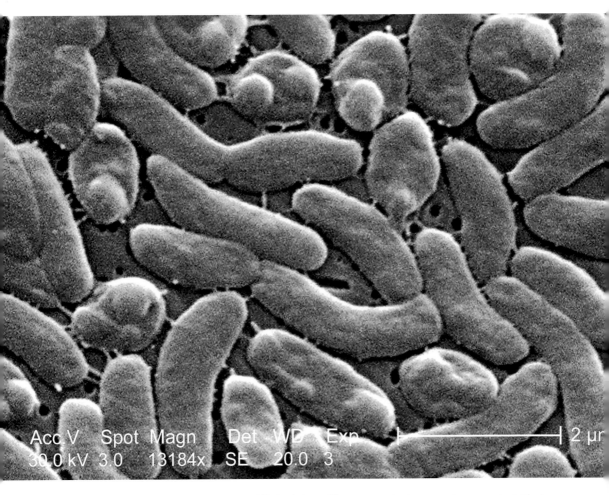

Acc V   Spot Magn   Det WD   Exp
30.0 kV 3.0   13184x   SE   20.0 3   2 µr

FIGURE 3.5 *Vibrio cholerae.* The green color is computer generated. Photo courtesy CDC.

returned from India loaded with trade goods and some unfortunate sailors suffering from cholera. Its bilge was filled with cholera-laced water the ship had taken on in the Bay of Bengal. After its arrival home, there were two cholera outbreaks. The first was in 1831 in Sunderland, near the port city of Liverpool, most likely initiated by the sick sailors, and the second was in 1832 in London. When the ship docked in London and emptied its bilge water, the bacteria successfully invaded and colonized the nutrient-rich Thames estuarial ecosystem. From that point on, outbreaks became a regular London occurrence, particularly in the districts nearest the river.

By the mid-1850s, the germ theory of disease was gaining traction. John Snow, an enlightened London medical practitioner, was thoroughly convinced cholera was caused by a germ. Snow was keen on the origins of infectious diseases and closely followed whatever literature was available. He speculated that cholera might even be water-borne and was committed to finding out what was going on. In the summer of 1854, during the height of a seasonal outbreak, he began systematically tracking where the sick lived and which of the several companies specializing in water services had supplied their homes with drinking water. His investigations included thousands of customers. He eventually narrowed his area of study down to Broad Street, where most of the cholera deaths that year had occurred, and determined that whoever became sick used the same pump for their daily water supply. Snow summarized his findings: "The result of the inquiry then was that there had been no particular outbreak or prevalence of cholera in this part of London except among the persons who were in the habit of drinking the water of the above-mentioned (Broad Street) pump-well."[12]

Confident that he had now found the source of the contaminated water, Snow removed the pump handle, preventing its further use. However, the spring rains were long over, and the flow of the Thames had slowed down, allowing London's estuary to regain its normal level of salinity. The zooplankton had reproduced, and the cholera bacteria, in response to the higher level of saline, had transformed back into its noninfectious resting stage that now lay dormant on the bottom of the riverbed. In the following years, if the spring rains were heavy enough, it would again dilute the salinity of the estuary, triggering a new cycle of zooplankton reproduction.[13] Another cholera epidemic would soon follow.

Snow's seminal studies became the foundation for the modern science of epidemiology, the keystone of today's public health practice. The discovery that cholera was indeed the offending waterborne microbe, first by Filippo Pacini in Italy in 1854 (albeit little noted) and then by Robert Koch in India in 1883 (his findings were widely publicized), started a global surge of public health initiatives to secure a safe water supply. By the beginning of the 1900s, as a direct result of Snow's investigations, cities throughout most of Europe and the United States emphasized sanitation as the best means of slowing down the rate of cholera outbreaks and improving the public's health. As an unexpected bonus, the incidence and prevalence of other water-borne diseases—typhoid fever, bacterial and amoebic dysentery, and giardiasis—also decreased. Infant mortality rates dropped, and life expectancy throughout the Western Hemisphere increased measurably, at least among the majority of middle-class inhabitants of most cities.

The Invention of Sanitation

In the United States, the health of nearly half of the newly founded country's population was less than adequate as the result of exposure to a different kind of pathogen. But like cholera, the problem was resolved with the application of basic sanitation, establishing once and for all the power of public health to ensure the general well-being of millions of people.

The American Civil War officially ended on April 9, 1865. But the years following that bloody conflict were filled with hardship that rivaled the war itself. The South's economy was slow to recover. The conventional wisdom, particularly among most northerners, was that the southern whites were just plain lazy. Eventually, the real cause of "Southern laziness" was determined. The majority of rural white southerners were suffering from anemia. The Rockefeller Sanitary Commission was established, and it identified hookworm (see fig. 3.6), a blood-sucking parasitic

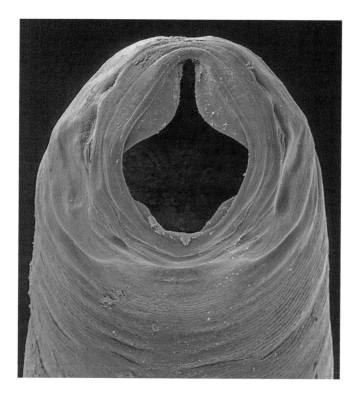

FIGURE 3.6 Head of *Necator americanus* (hookworm). Photo courtesy David Scharf.

roundworm, as the real reason people could not put in a full day's work.[14] Subsequently, the control and partial eradication of hookworm infection, a geohelminth transmitted by fecal contamination of the soil, was facilitated by the invention of the outhouse, grassroots public health education, and the eventual urbanization of the South.

In England, the invention of the flush toilet, coupled with efficient wastewater treatment methods, enabled communities to upgrade their liquid wastewater control schemes, further improving the public's health.[15] This proactive public health advocacy benefitted the urban community at large and allowed them the luxury of expecting a better life. In many instances, affluent communities were more than willing to financially support various health initiatives by advocating for increases in local taxes. In the

United States, federal standards were established regarding the construction of municipal wastewater treatment facilities and safe sewage systems.

The establishment of science-based public health regulations and enforcement of city-wide laws related to them became a reliable, effective strategy to protect urban populations from a wide range of illnesses. Still, there are instances when even this degree of oversight failed. One notable breach of security occurred in Milwaukee, Wisconsin, in the spring of 1993. A portion of that city's wastewater treatment facility was under repair and resulted in the world's single largest diarrheal disease outbreak in modern times. An unanticipated spring thaw and the unfortunate location of an effluent pipe from a nearby slaughterhouse adjacent to the intake pipe for the section of the water treatment plant being worked on created the epidemic. *Cryptosporidium parvum* (see fig. 3.7) was the offending pathogen, a ubiquitous waterborne

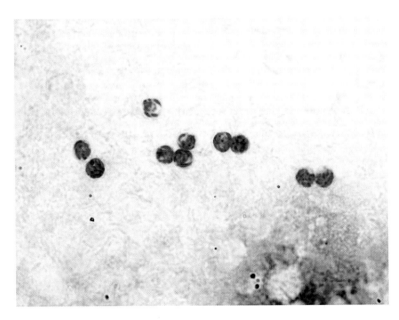

FIGURE 3.7 Oocysts (red-stained spheres) of *Cryptosporidium parvum*. Photo by author

parasite of domestic cattle that occasionally also infects people. Regrettably, over three hundred Milwaukeeans who were infected, and who also unknowingly harbored the HIV virus, died from overwhelming infection.[16]

The definition of public health gradually expanded to include air quality monitoring; regular inspections of food-processing facilities, abattoirs, and restaurants; and numerous educational outreach programs, including maternal and neonatal health care, family planning, and nutrition. Solid and liquid waste management began to be controlled by sanitation departments. Countries established national Departments of Public Health, and universities and colleges began training students in the new science of public health. In 1899, the London School of Hygiene and Tropical Medicine became the world's first school of public health. In 1918, Johns Hopkins University established the first school of public health in the United States. As of 2023, there were 149 accredited graduate schools of public health in the country.

HAS SANITATION BECOME A LUXURY?

Despite much progress, too many cities worldwide still lack an effective integrated public health system. Many are in less developed countries (LDC) in South and Southeast Asia, but Central and South America and Africa also have their share of underserved urban populations. Infant mortality rates (see table 3.1) and average life expectancy are two indicators of how a country deals with health issues. In the majority of LDCs, infant mortality rates are significantly higher compared with developed countries.

Many of the disenfranchised living in LDCs have little or no access to health services. Substandard housing, lack of access to public transportation, food deserts, limited or no green spaces in poor neighborhoods, and locating pollution-producing municipal services such as wastewater treatment plants and bus terminals in poorer neighborhoods are some of the

TABLE 3.1.

Infant mortality rates for the twenty least developed countries

Country	Mortality Rate	2022 Population
Afghanistan	110.6	40,754,388
Somalia	94.8	16,841,795
Central African Republic	86.3	5,016,678
Guinea-Bissau	85.7	2,063,367
Chad	85.4	17,413,580
Niger	81.1	26,083,660
Burkina Faso	72.2	22,102,838
Nigeria	69.8	216,746,934
Mali	69.5	21,473,764
Sierra Leone	68.4	8,306,436
DR Congo	68.2	95,240,792
Angola	67.6	35,027,343
Mozambique	65.9	33,089,461
Equatorial Guinea	65.2	1,496,662
South Sudan	62.8	11,618,511
Zambia	61.1	19,470,234
Gambia	60.2	2,558,482
Comoros	60.0	907,419
Burundi	58.8	12,624,840
Uganda	56.1	48,432,863

Note: Per 1,000 live births.
Source: World Bank, "Mortality rate, infant (per 1,000 live births)," https://data.worldbank.org/indicator/SP.DYN.IMRT.IN.

environmental injustice issues that continue to affect billions of people worldwide.

Sources of drinking water for municipal systems include surface waters (lakes, reservoirs, rivers), groundwater from subterranean aquifers (artesian and pump-assisted), and saltwater that is made drinkable using reverse osmosis. Many countries acquire their drinking water from the ocean by desalination (i.e., distilling). In. contrast, the British island territory of Bermuda gets all its freshwater by harvesting and storing rainwater.

New York City accesses its drinking water from reservoirs upstate. The city's Department of Environmental Conservation states, "New York City gets its drinking water from 19 reservoirs and three controlled lakes spread across a nearly 2,000-square-mile watershed. The watershed is located upstate in portions of the Hudson Valley and Catskill Mountains that are as far as 125 miles north of the City" (see fig. 3.8).

These reservoirs deliver 1.2 billion gallons of water daily via a large-bore underground tunnel system to reservoirs closer to the city. From there, water travels through another tunnel to fill a central reservoir in Manhattan's Central Park, where it is then distributed to every building. Water pressure forces it to rise to the height of six stories. Taller buildings have water towers that are filled by pumping, and then the water travels down to all the floors below.

New York does not filter most of its drinking water (880 million gallons) since its watershed forests serve as an effective buffer, protecting its residents against exposure to contaminants emanating from surrounding farmland and rural communities. The water that comes from the Croton Reservoir (320 million gallons) is filtered to remove contaminants from more densely settled areas that lie closer to the city limits. New York then treats the combined water with chlorine and UV light to kill off most offending pathogens prior to distributing it to the public at large. Despite this vigilant (and expensive) public health oversight, the system is not free of concerns. Things that can adversely affect a surface water supply include prolonged droughts, contamination from accidental spills, fish kills due to high summertime temperatures, and severe floods that transport contaminants from adjacent developed land.

Groundwater is not as common a resource for cities, but roughly 90 percent of rural and suburban communities rely

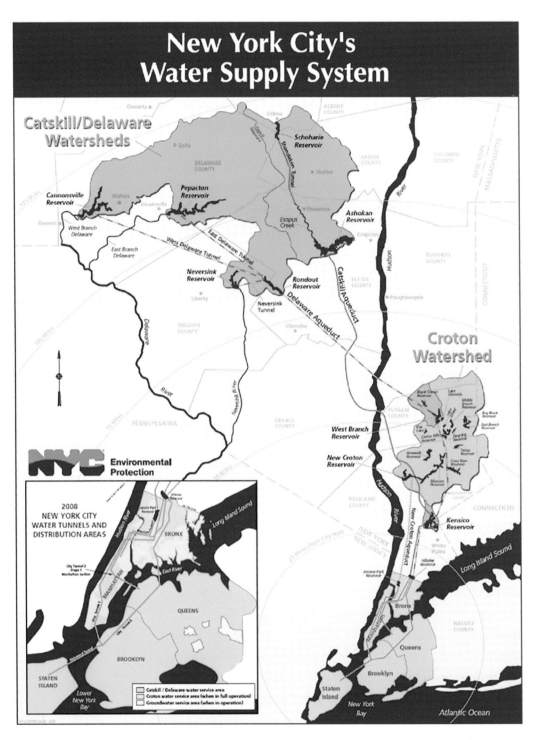

FIGURE 3.8 Map of New York City's reservoir system. Courtesy NYC Department of Environmental Protection.

on it for all their water needs.[17] Depending on how deep the wells are drilled, groundwater can present health risks related to both naturally occurring (e.g., arsenic, uranium, lead, chromium) and anthropogenic sources of noxious chemicals (e.g., volatile organics, formaldehyde, vinyl chloride, and other industrial wastes).

Where public health oversight is inadequate or absent, drinking contaminated groundwater has created serious health problems. Bangladesh's water, which is collected from its rivers and aquifers, emanates from the Himalayas. Shortly after the country's founding in 1972, shallow wells were dug as part of a World Health Organization–initiated public health upgrade to eliminate the use of surface water for drinking, then a major source for diarrheal diseases. Infection rates for a wide variety of waterborne infectious diseases plummeted and everything appeared to be on course to significantly raise the health status of millions of people. But daily consumption of well water eventually compromised the health of those same people. They began to suffer from numerous illnesses, including various cancers, cardiovascular disease, and impaired cognitive development, resulting in a shortened life span. It was then discovered that the water in many of the wells contained an unusually high level of arsenic. Many of the most heavily impacted areas were helped by the installation of deeper community wells that avoided the arsenic-laden shallow aquifers.[18]

Lead is another element that continues to plague the modern world, especially within the built environment. Lead poisoning has been officially recognized as a medical condition since the 1800s, but it was not until the 1960s that serious public health concerns raised enough public awareness to initiate rigorous scientific studies on lead and its relationship to diseases associated with exposure to high levels of it. Data rapidly accumulated leading to the eventual elimination of the use of lead in gasoline, paint, and pipes. Most of those studies confirmed the detrimental effects of continued exposure to even small amounts of lead.

The Centers for Disease Control and Prevention lists the following areas of concern:

Exposure to lead can seriously harm a child's health and cause well-documented adverse effects: damage to the brain and nervous system, slowed growth and development, learning and behavior problems, and hearing and speech problems.

Unfortunately, plumbing and house paint remain the main sources of lead in cities and suburban communities, especially in older residential buildings commonly found within poorer neighborhoods. Since pipes made of lead are easy to manipulate, never rust, and are cheap to produce, an enormous amount of lead piping was installed before public health regulations banned it.

Mothers who live in lead-contaminated environments have elevated levels of lead in their breast milk that then can be ingested by their infants. Children who grow up in such situations show reduced growth rates and impaired intellectual skills. These effects are long-lasting and may even be permanent.

Because of the lead plumbing poorer neighborhoods have the highest rates of lead in their drinking water. The only solution is to replace all of those pipes, but that is an expensive and time-consuming strategy. Programs to implement their replacement have been slow to start, again reflecting the differences between the haves and have-nots. Flint, Michigan, is a prime example of this kind of egregious and discriminatory governing behavior. In the new city, this kind of health risk will be excluded by design.

Drinking water is not the only source of exposure to ground contaminants. Plants will take up almost any soluble, low-molecular-weight compound. In fact, plants are frequently used to remediate land that has been rendered unusable due to the dumping of industrial wastes. That is why plants are sometimes referred to as living machines.[19] Crops irrigated with wastewater, a common practice in many parts of the world, can accumulate heavy metals, pesticides, and herbicides.[20] Rice grown in many places contains high levels of arsenic.[21]

Getting safe drinking water to the city serves as a focal point for learning how the problems and solutions to managing the public's health have evolved over the last ten thousand years. Planning a new city takes these lessons learned into account and will employ different strategies for ensuring sustainable, long-term resilience regarding drinking water security. The safest and easiest way to achieve this goal is to create cities that are independent of centralized water delivery systems. Getting off the water grid makes sense provided that proven, robust technologies are applied. Fortunately, there are many options to choose from for securing and safely storing drinking water.

I propose a circular economy of water use, where every building will have the capacity to capture and store water from a variety of sources. For example, harvesting water directly from the air is a viable option that simplifies the acquisition of a reliable and safe water supply. Each structure will also have the technology to recycle all wastewater back to drinking water standards.[22] At the municipal level, this is already an ongoing practice in many places.[23]

Freeing up cities from relying on the natural landscape for their water needs allows for the rewilding of numerous fragmented habitats. When this day comes, the natural world will once again be able to maintain ecological equilibrium without interference from us, and we can continue our evolutionary journey toward a brighter future.

THE CHANGING REALITY OF WATER RESOURCES

Having enough water for our crops is another daunting problem because of rapid climate change (RCC). Globally, over the last several hundred years, we have built millions of dams (see

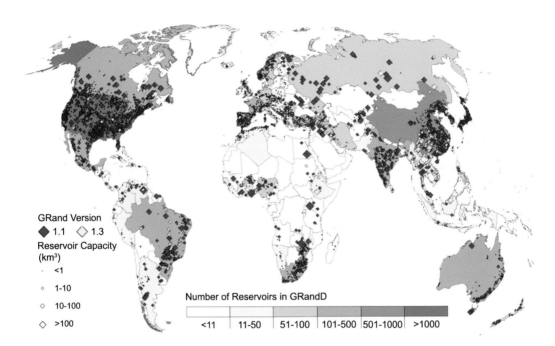

GRand Version
◆ 1.1   ◇ 1.3
Reservoir Capacity
(km³)
· <1
◦ 1-10
◇ 10-100
◇ >100

Number of Reservoirs in GRandD

| <11 | 11-50 | 51-100 | 101-500 | 501-1000 | >1000 |

FIGURE 3.9 Global distribution of dams. Courtesy Global Dam Watch.

fig. 3.9), eighty-four thousand within the United States alone, whose main purpose is creating reservoirs for irrigation.[24] Erecting dams across rivers has dramatically altered the ecology of countless waterways and their watersheds—and usually not for the better.

A case in point is the system of dams along the Colorado River; there are fifteen on its main stem and hundreds more on its tributaries. Over the last ten years, the snowpack in the Rocky Mountain watersheds of western Colorado has been well below average. That means less water is delivered to western Arizona, southern Nevada, and California. The bulk of California's share irrigates crops that supply a large portion of the United States with fresh produce. The recent failure of many crops throughout the Central Valley of California due to drought attests to the ephemeral nature of that once-reliable source of freshwater.[25]

Agriculture in Colorado has also suffered from that same reduced snowpack, forcing it to renegotiate its water agreements with Arizona, New Mexico, Nevada, and California. Reducing water to an already water-starved $50 billion-a-year agricultural system in California will have serious economic consequences for the entire country and could even force its collapse if the drought continues for another five to ten years.

Worldwide, more than 70 percent of irrigated farmland and nearly 50 percent of cities experience water shortages at some time during the year.[26] Drought has also made many hydroelectric dams less efficient than when first constructed. Collectively, the hydroelectric dams on the Colorado supply energy to 780,000 people. Reduced water pressure in reservoirs behind those dams directly affects the amount of electricity each dam can generate.[27] At the time of this writing, Lake Powell had lost 77 percent of its water to the drought. If the drought worsens, its adverse effects on all life within that zone will become even more pronounced, especially considering this climate event has been ongoing for the past 1,200 years. Williams, Cook, and Smerdon state in their abstract, "A previous reconstruction back to 800 CE indicated that the 2000–2018 soil moisture deficit in southwestern North America was exceeded during one megadrought in the late-1500s. Here, we show that after exceptional drought severity in 2021, 19 percent of which is attributable to anthropogenic climate trends, 2000–2021 was the driest 22-yr period since at least 800. This drought will very likely persist through 2022, matching the duration of the late 1500s megadrought."[28]

During periods of protracted drought, a city has only a few options for continuing to provide drinking water to its residents. It must either reduce the daily per capita amount consumed by prohibiting nonessential use or rely on brackish or saltwater sources, but converting those other sources to drinking water quality is expensive. Instead, many cities are now investing in systems designed to recycle gray water, taking advantage of new "toilet-to-tap" water treatment technologies.[29]

When the rain finally does come, the result is often flooding that exceeds the capacity of wastewater treatment plants, especially in cities like New York that have combined sewage systems. Over the ensuing weeks, floods create opportunities for microbial pathogens, mostly diarrheal disease-causing agents, to enter the built environment.

Historically, the invention of public health grew out of a need to manage liquid waste. Creating wastewater treatment plants isolated urban populations from coming into direct contact with untreated urine and feces. As cities grew, the demand for drinking water increased, placing added pressure on treatment plants to remain true to their promise of maintaining the highest level of public health standards. Eventually, in many instances, population growth overwhelmed outdated water supplies, creating serious health consequences.

CREATING A CIRCULAR WATER ECONOMY

For the new city to exist in harmony with its surrounding landscape, it must solve its waste management issues in ways that do not strain the environment. Each building will be responsible for the management of its liquid and solid waste streams. Practical, scalable solid waste-to-energy technologies that are commercially available take advantage of the process of high-temperature incineration; plasma arc gasification (PAG), for example, eliminates the need for landfills while generating heat that is converted into electricity through a steam turbine system. PAG has been successfully applied at the municipal level, especially in China, Japan, and Indonesia.

Wastewater treatment plants that remediate blackwater to drinking water quality are gaining in popularity at the municipal level, and some have garnered support from the Bill & Melinda Gates Foundation. For example, Orange County, California,

FIGURE 3.10 Water purification system for Orange County, California.
Photos courtesy of Orange County Water District.

converts all its wastewater to potable standards (see fig. 3.10) and
uses it to recharge the surrounding aquifers.[30] Drinking water is
then extracted from groundwater. California, Texas, Australia,
Singapore, Namibia, South Africa, Kuwait, Belgium, and the
United Kingdom are in the process or have already converted
some of their wastewater treatment plants to accommodate the
full recovery of water for reuse.[31]

Large sums of federal as well as local township funds are ded-
icated each year to an extravagant sanitation task—desludging
blackwater and dispatching the treated gray water into estuaries
and rivers (see table 3.2). It illustrates the extent to which we have
historically undervalued this precious resource.

TABLE 3.2.

Potential Water Harvesting for the 21 Most Populated Cities

City	Population (millions)	Square Miles	Annual Rainfall (inches)	Gallons Harvested (billions)
Tokyo	37.3	847	60	882
Delhi	31.2	573	31	300
Shanghai	27.7	2,448	47	2,127
São Paulo	22.2	587	53	510
Mexico City	21.9	573	41	200
Dhaka	21.7	118	73	154
Cairo	21.3	1,191	1	20
Beijing	20.8	6,336	22	2,202
Mumbai	20.6	233	89	350
Osaka	19.1	87	58	76
Karachi	16.4	1,459	7	178
Chongqing	16.3	31,776	45	27,661
Istanbul	15.4	2,063	33	1,057
Buenos Aires	15.2	78	44	54
Kolkata	14.9	80	63	83
Kinshasa	14.9	3,848	43	2,677
Lagos	14.8	452	70	550
Manila	14.1	16	66	8
Tianjin	13.7	4,541	24	1,700
Guangzhou	13.6	2,870	68	3,570

Source: Author, from USGS data.

There are only a few examples of individual commercial build-ings or apartment complexes that recycle all their liquid waste. In 2015, San Francisco passed a law requiring all newly constructed large buildings to recycle all liquid wastewater. One company, Epic Cleantec, used a grant from the Bill & Melinda Gates Foun-dation to install a state-of-the-art treatment system in the 61-story Salesforce building (see figs. 3.11 and 3.12). In the near future, it is expected that many more of these systems will come online.

FIGURE 3.11 The Salesforce building, San Francisco. Creative Commons Author Dead.rabbitt

FIGURE 3.12 The Salesforce building in San Francisco recycles all its water into drinking water quality. New York Times/Redux Pictures

GETTING OFF THE WATER GRID

The new city should be independent of centralized water treatment facilities. Instead, it will take advantage of a wide variety of atmospheric water harvesting (AWH) and storage technologies that emulate ecosystems, which are more sustainable. At the same time, buildings will be able to share their stored water reserves, emulating how trees function in a temperate zone forest.

Harvesting Water from the Air

Relying on surface water for all of a city's water needs is unsustainable. The chronic shortage of that resource has forced nearly every municipality to reevaluate their long-term water management strategies. The United States Geological Survey (USGS) studies the water cycle on a global scale and is an excellent resource for how cities manage their water supplies. The USGS has calculated what might be possible if a city were to capture its annual rainfall and use it, instead of surface water, for all its water needs:

> Consider for a moment how much rainwater some cities may receive during a year. For example, Atlanta, Georgia (population 488,800), averages about 45 inches of precipitation per year[, which works out to] 103.2 billion gallons of water. . . . The per capita water use is about 110 gallons per day or 40,150 gallons per year. Thus, the water from a year's precipitation, if it could be collected and stored without any loss, would supply the needs of about 2,574,000 people.

The two following case histories for Cape Town and Mexico City will reinforce the usefulness of rainwater harvesting (RWH).

Cape Town, South Africa, covers 154 square miles and depends on six reservoirs for its entire supply of freshwater. It receives twenty inches of rain on average each year, totaling fifty-three billion gallons, but none is collected. From 2015 to 2017, the city suffered one of its most severe droughts. In early 2018, Cape Town came very close to using up its entire water reserves. The situation became known as the "Day Zero" drought. Salvatore Pascale and his associates' summarized the conditions that led up to that water shortage:

> The Cape Town "Day Zero" drought was caused by an exceptional 3-y rainfall deficit. Through the use of a higher-resolution climate model, our analysis further constrains previous work showing that anthropogenic climate change made this event five to six times more likely relative to the early 20th

century. Furthermore, we provide a clear and well-supported mechanism for the increase in drought risk in [southwestern South Africa]. . . . A reduction in precipitation during the shoulder seasons is likely to be the cause of drought risk in southwestern South Africa in the 21st century. Overall, this study greatly increases our confidence in the projections of a drying SSA.[32]

Cape Town narrowly escaped Day Zero when the rains finally came in 2018, filling up all six reservoirs. The city then initiated three projects designed to bolster and expand options for acquiring freshwater. This change in policy added a small measure of resilience to their ability to provide an adequate amount of potable water to its residents. The first initiative was to access groundwater from several aquifers near the city limits, reducing their reliance on surface water. The Cape Town–based energy corporation, Crown Energy, supplied the metrics for the feasibility of accessing the nearby groundwater aquifers: "The Cape Flats aquifer, the Table Mountain aquifer and the Atlantis aquifer . . . can deliver, as per early estimates, 80, 40, and 30 megalitres per day respectively."[33]

The second initiative established four modestly sized desalination plants. The third was to construct a wastewater remediation facility. As of 2023, however, the city council had not yet identified RWH as a potential municipal activity, even though some companies, such as WET Technologies, specialize in selling and installing rainwater harvesting equipment and the city encourages its citizens to consider getting involved in harvesting rainwater. Neglecting to make use of even a small portion of fifty-three billion gallons of accessible water each year in a drought-prone area of the world is an obvious missed opportunity.

The greater metropolitan area of Mexico City, which covers 573 square miles, has experienced chronic droughts since its establishment in 1521, even though its annual rainfall averages forty-one inches or approximately 430 billion gallons. Unlike

Cape Town, it requires all new buildings to have the capacity for RWH.[34] In addition, numerous households have adopted the strategy and can now obtain most of their needed water from rooftop catchment facilities.

Other cities that suffer occasional chronic water shortages include Los Angeles, New York, San Francisco, London, Moscow, São Paulo, Bangalore, Beijing, Cairo, Jakarta, Istanbul, and Tokyo, but only Bangalore has mandated RWH. Countries that lead the way in developing and applying rainwater harvesting systems include India, Australia, Brazil, Germany, China, and Singapore. In every case where RWH is supported, the reason given is the same: to address changes in the water cycle due to rapid climate change.

What follows is an overview of the most promising methods of accessing water from the air to help the new city become independent of centralized municipal water supplies. The unintended positive consequences of doing so are staggering, especially for the world's terrestrial and freshwater ecosystems. Ecologists are universally optimistic at the prospect of more enlightened, resource-conscious urban behavior regarding water usage. In this regard, harvesting water from the air qualifies as one of the most beneficial changes for both city dwellers and wildlife.

The new city will incorporate multiple technologies for collecting water from the atmosphere. As well, numerous recently developed rainwater harvesting inventions and innovations promise to enable any municipality, regardless of location, to collect and store enough water to comfortably exceed its annual consumption. Wastewater will no longer be discarded into the environment.

Geography plays a central role in what kind of water harvesting technologies a city might employ. Annual precipitation and ambient temperature profiles contribute to determining the amount of water available for every ecological zone. Recall that the majority of cities throughout the world access surface water for all of their needs. Dams play a central role in creating reservoirs for this purpose. In the past, the majority of the world's reservoirs were more than adequate to meet demands. Hyperurbanization and RCC are now responsible for chronic water shortages. New ways to secure a reliable, sustainable freshwater supply must be implemented if cities are to survive and thrive. The most efficient strategies for harvesting water from the air are rainwater harvesting, harvesting water from fog and dew, and dehumidification. This article by Christopher McFadden provides a thorough overview of the subject.[35]

Rainwater Harvesting

As the Earth continues to warm, annual precipitation expectations will shift, both with respect to the amount of rain and the yearly timing of rain events. This will have profound effects on the availability of water for the entire planet. Due to the limitations of all climate models, however, accurately predicting where the greatest changes will occur and what their effects will be, is still not possible. Despite their discrepancies, all models agree about one thing; a warmer climate is an eventuality.[36] The inclusion of RWH into so many countries' master plans for dealing with this resource is why we now have such a robust menu of technologies from which to choose.

Many ancient cultures practiced it in some form. The most straightforward approach was to create reservoirs by damming rivers or diverting a portion of one into a holding tank of sorts. But unlike these early attempts at establishing an abundant source of freshwater, modern applications are scalable, ranging from single-family dwellings to apartment complexes and office buildings.

One of the most successful and striking systems exists on the island of Bermuda. There, the roofs are specially designed (see fig. 3.13) to channel rainwater into underground cisterns. Bermuda is one of the best modern examples of how people

FIGURE 3.13 Grooved roof of a typical home in Bermuda. Rainwater is channeled to an underground cistern where it is filtered and stored. ShareAlike 3.0 Unported (CC BY-SA 3.0)

rally around the concept of rainwater harvesting when nature dictates the terms under which they must live. With a population of nearly sixty-four thousand, everyone is engaged in this activity. Because Bermuda has no surface water or groundwater sources, relying on rainstorms for their water needs is their only option.

Meticulously maintaining clean roofs reduces the chances of unwanted contaminants entering their water supply. This need for extreme outdoor hygiene has given Bermuda its distinctive personality of visually pleasing contrasts—sparkling white roofs and pastel-colored homes, making it an attractive international destination for tourists. The island's biggest problem is how to deal with its wastewater. Most of its septic tanks are filled to the brim, and there is little room left to install more. One practical solution is to recycle all wastewater, and it appears that plan is about to go forward.[37]

All buildings in the new city will be designed to employ RWH wherever the climate permits. A minimum of twenty inches of rain per year will suffice for populations that do not exceed one million. Many versions of RWH devices are employed throughout the world.

A set of ten to twenty interconnected buildings could collectively supply a common underground reservoir with rainwater. Each contributing building would access water from it as needed. Every structure in the new city must guarantee the purity of the water it is consuming, so the installation of easy-to-maintain purification systems, such as filtration and UV sterilization devices, would be essential for this arrangement to function smoothly and safely. For cities that cannot build below ground, holding tanks would be integrated into the architecture of each structure allowing it to collect and store water.

Imagine upside-down umbrella-like devices stored in cylindrical tubes on all four corners of each structure, and the tubes are next to water storage cylinders. They would emerge and unfold at the beginning of a rain event and perhaps even lock their edges together by electromagnets, making the collection area more expansive. The water would be funneled through the cylinders into the underground reservoirs. At the end of the storm, the umbrellas would dry out, detach from each other, and fold back into their respective storage tubes. If even one-tenth of New York City's annual rainfall could be captured and stored this way, it would greatly reduce or perhaps even eliminate any possibility of a water shortage.

Mega-metropolises such as Beijing, Chongqing, Shanghai, and São Paulo will need a robust mix of solutions to their water management, with RWH at the forefront.

Roof design will undoubtedly figure prominently in the new city, paying homage to Bermuda, but employing completely different geometric configurations to maximize the amount of water captured.

Harvesting Water from Fog and Dew

Numerous locations along the coastlines of most continents are ideal for harvesting water from fog (see figs. 3.14 and 3.15). In the last five years, many countries have focused on these opportunities as a means of augmenting water supplies for urban and rural communities—the Canary Islands, Chile, Colombia, Eritrea, Ethiopia, Guatemala, Israel, Morocco, Namibia, Oman, Peru, and South Africa.[38] The amount of water contained in a given volume of fog is often large enough to meet the daily demands of many coastal communities, although the actual amount that can be harvested using current methods is still inadequate for large cities.

FIGURE 3.14 Fog harvesting device designed for a small town. Creative Commons photographer Nicole Suffie

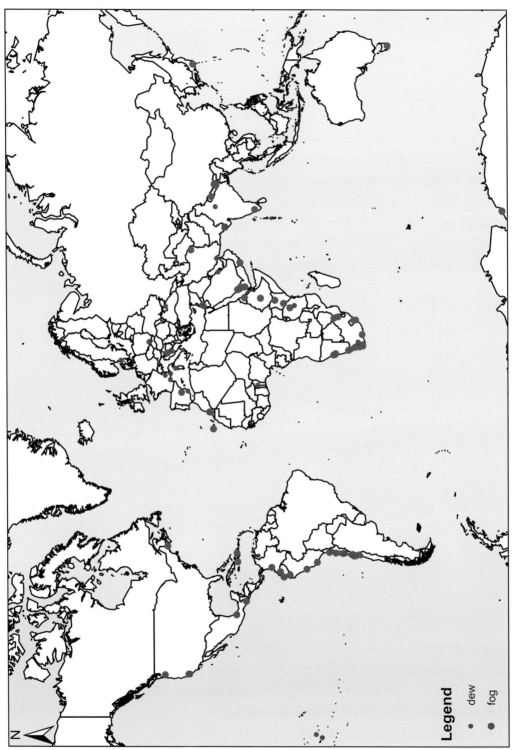

FIGURE 3.15 Global locations ideal for harvesting water from fog. Freemaps.

Legend

· dew

● fog

Dev Gurera and Bharat Bhushan study fog-harvesting technologies. They researched and then deconstructed natural systems, learning how desert life in fog-rich coastal zones acquired water (see fig. 3.16), particularly the more successful plants and insects.[39] Their investigations revealed that those life forms were able to extract moisture almost out of thin air because of the microscopic topographies of their outer surfaces. The researchers then replicated those modified geometric features of various microscopic surfaces of fog-harvesting cacti and insects using 3D printing. They tested the ability of each sheet of plastic-coated 3D images for their efficiency at harvesting water from laboratory-generated fog. Many designs performed well under

FIGURE 3.16 The Darkling beetle harvesting water from the morning fog on the coastal side of a Namibian sand dune. Solvin Zankl/NPI/Miden Pictures

controlled conditions, providing yet another example of how nature can inspire the creation of a new technology that can then be applied to the built environment.

Another approach to the design of water collection devices that interface with fog was exploited by Joanna Knapczyk-Korczak and colleagues.[40] They incorporated randomly oriented electrospun polyvinylidiene fluoride fibers into Raschel mesh (a water-harvesting, cloth-like material) and improved water collection rates over 300 percent.

These findings, and similar studies, strongly support the concept that scalable, highly efficient fog-harvesting devices based on natural design principles will likely be commercially available over the next five years. However, ocean fogs are laden with minute salt particles that serve as condensation points during fog formation. Water collected from fogs along coastal zones must be desalinated before consumption. Collecting water from dew by employing various dehumidification technologies has also become practical, greatly expanding the areas where water harvesting can occur.[41]

Applying the findings from the studies described above to our new city appears to be straightforward. For example, the facade of all buildings located in fog-rich areas could consist of a thin, nonflammable polymer onto which the most efficient microscopic surface features of resilient desert-dwelling life forms have been incorporated. In this way, every building, in addition to having RWH capacity, could also routinely collect water from fog. This hybrid strategy should enable any new city, regardless of location, to capture and store a sustainable freshwater supply and without harming the surrounding landscape.

Dehumidification

Two other approaches for extracting water from the environment are dehumidification and desalination. Dehumidification depends on establishing a temperature gradient between a device, in which a solution colder than the ambient atmosphere is maintained and circulated, and the surrounding air. Water droplets form around specially designed fixtures, coalesce, and trickle down into a collection vessel. Drawbacks to dehumidification include its requirement for advanced technologies and energy input to operate the pump that facilitates the circulation of the cold fluid and a chiller that maintains that fluid's lower temperature. The use of renewable power sources to cool the circulating fluid should enable this technology to become better established over the next few years.

Collecting water that condenses from air conditioners has been reported. In one study conducted in Brazil, various sizes of air conditioners (9,000 BTU, 12,000 BTU, and 24,000 BTU) were analyzed for their hourly output of condensate.[42] The two smallest units yielded 3 L/hour, while the 24,000 BTU unit produced 4.1 L/hour over nine hours.

DESALINATION

Desalination does not involve capturing water from the air, but it should be considered alongside the other technologies that enable a city to go off the centralized water grid. Desalination is gaining popularity at the municipal level as a practical means of obtaining large quantities of freshwater from the oceans (see fig. 3.17).[43]

Currently, about 1 percent of Earth's eight billion people rely on desalination for some or most of their water needs. Creating potable water from seawater involves distillation or reverse osmosis (RO). Both technologies are energy-intensive (the average is 3 kWh/m^3). Another drawback is that many distillation facilities use fossil fuels to boil seawater.

Applied research on renewable energy sources, particularly wind and solar, to supply power for RO facilities continues to show improving efficiency. Desalination has the possibility of becoming a carbon-neutral technology within the next ten years.

Desalination plants worldwide

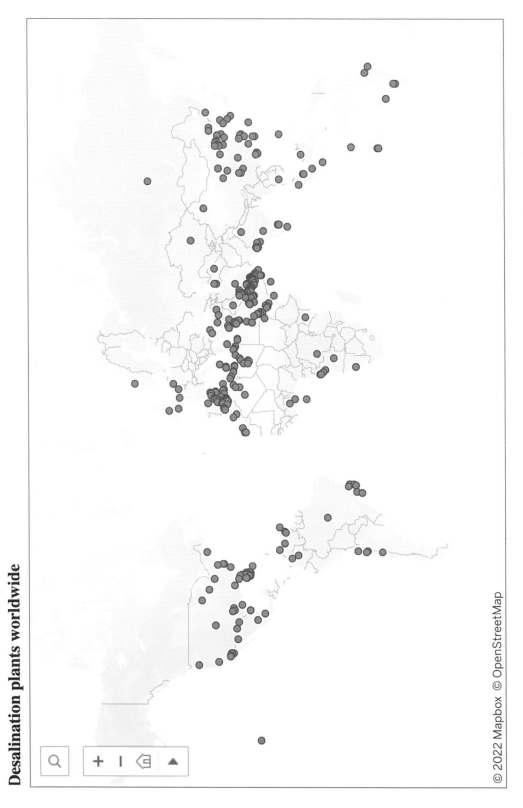

© 2022 Mapbox © OpenStreetMap

FIGURE 3.17 Global distribution of desalination plants. Green dots show where salt deposits contaminate the groundwater supply. openstreetmap.org

New coastal cities will undoubtedly benefit from this research. Environmental concerns are focused on what to do with the hypersaline wastewater (effluent) produced by both methods. Suggestions include "mining" the effluent for rare earth elements and exotic metals commonly found in trace amounts in seawater or using it in various configurations of cooling systems.

Not having to rely on surface water or groundwater to obtain freshwater will dramatically improve the odds of helping to restore damaged ecosystems, and at the same time, reduce the number of outbreaks of infectious diseases spread by contaminated water. Securing our own water supply from the air makes us independent of the natural landscape and allows us to live alongside it without encroaching on the very cycles of life that foster the health of every living thing, including us.

4

Pillar Four
Renewable Energy

Solar architecture is not about fashion,
but rather about survival.
—SIR NORMAN FOSTER

THE VALUE OF RENEWABLE ENERGY

E nergy, the fourth pillar of urban sustainability, is the most problematic. Nothing happens without energy. Energy and matter govern the behavior of the universe—from the trigger that started the Big Bang to the subatomic forces created by that cataclysmic event. Nearly six hundred species of animals live off the heat and nutrients brought to them by hydrothermal vents (see fig. 4.1). That energy is generated by thermonuclear reactions deep inside Earth's mantle. In contrast, the rest of life on the planet relies on solar energy for all its needs. Put another way, humanity lives because the sun shines.

City buildings consume 40 percent of the total global power budget.[1] Most of that energy is generated using fossil fuels. The new city will be powered by 100 percent renewable energy. Among the handful of commercially viable technologies, solar power leads the way to a cleaner future. Growth of the solar power industry has been so rapid over the last five years (see fig. 4.2) that in just a few years entire cities will be able to leave the municipal energy grid. The United States Department of Energy estimates

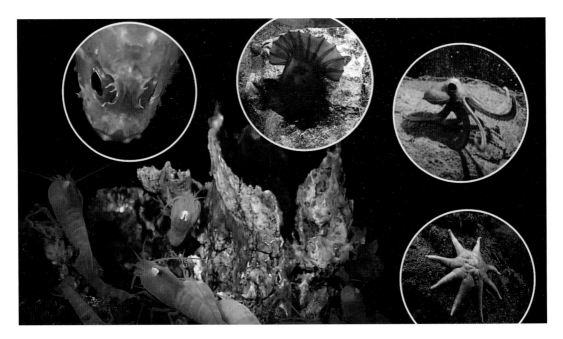

FIGURE 4.1 Thermal vents with associated ocean life. Courtesy NOAA

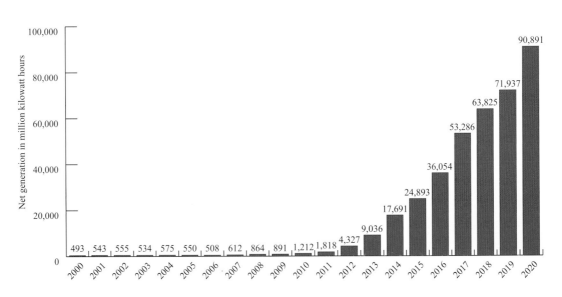

FIGURE 4.2 Annual growth of the solar power industry.

that "the amount of sunlight that strikes the earth's surface in an hour and a half is enough to handle the entire world's energy consumption for a full year."[2]

Trees rely on photons (solar radiation) for the generation of chemical energy. Their leaves convert that energy through photosynthesis. A portion of it is used to synthesize a wide variety of sugars, including cellulose. Along with lignin, they are the building blocks of the tree itself. Solar panels also rely on photons from the sun; the photons displace electrons from a thin layer of crystalline silicon, which are then harvested for the production of electrical energy. While trees use sunlight to make storable energy in the form of sugars, we capture sunlight and use it to run a wide variety of devices. The biomimic connection lies in the fact that both trees and solar panels capture photons. We are still working on how to efficiently store electrical energy for long periods of time after we generate it.

Why haven't we been able to generate enough solar energy to benefit everyone? It's not for lack of trying, and perhaps in the near future we will achieve that goal. Electricity generated by silicon-based photovoltaic (PV) panels and wind turbines accounts for 10 percent of the total energy output for the United States.[3] If their growth continues, both of these technologies could replace fossil fuels as energy sources.

A state-of-the-art PV panel converts 23.6 percent of the incoming photons that strike its surface into electrons.[4] As a result, their commercial production has grown into a robust, profitable industry and is very competitive with carbon-based energy resources.

But despite steady progress in converting all power generation to clean energy sources, we continue to consume huge quantities of fossil fuels. "Business as usual until it fails" is the ethic of the short-term capitalist, and no amount of incriminating climate science data seems to change that mindset. Not even the fact that the World Health Organization conservatively states that nearly four million people die prematurely each year due to polluted air, mostly caused by the burning of fossil fuels. So be it. Let the free market and the climate determine which energy sources will power the future. I believe the answer is obvious.

When the world eventually does convert to renewable energy, rapid climate change (RCC) will slow down, even if we fail in the end to restore a large portion of the world's damaged forests. The new city is about achieving balance and harmony with the natural world. Relying solely on renewable energy sources is the only logical, ethical, and economical way to realize that goal.

Applying the most promising of the numerous technologies in renewable energy generation (REG) to the built environment is currently expensive, but at least it is possible. I will now summarize the progress made in the development of applied REG and how it might be incorporated into the new city.

RENEWABLE ENERGY GENERATION

Photovoltaic cell

While the amount of sunlight varies with latitude and the season, every city has the capacity to capture and use some solar energy regardless of its location. The invention of the solar cell (or PV cell) officially began thusly: "PV technology was born [in the United States] when Daryl Chapin, Calvin Fuller and Gerald Pearson develop the silicon PV cell at Bell Labs in 1954—the first solar cell capable of absorbing and converting enough of the sun's energy into power to run everyday electrical equipment. Bell Telephone Laboratories produced a silicon solar cell with 4 percent efficiency and later achieved 11 percent efficiency."[5]

But nothing happens in isolation, especially in science. Other important discoveries and insights from applied physics and material sciences contributed significantly to the development of the PV cell.[6]

Work on the design and chemical composition of PVs since 1954 has focused on increasing their efficiency in capturing photons. Witness the worldwide abundance of solar panels on roofs of homes, telephone poles, empty fields, and parking lots (see figs. 4.3 to 4.6). As of 2020, the estimated amount of their global capacity was 844 terawatts.[7] PV cells can be purchased in many colors, expanding the range of application to the exterior of new housing, commercial buildings, and municipal structures such as bridges.

Using PV cells in the new city is limited to rooftops and some exterior cladding strategies. The Powerhouse Telemark building in Norway is a good example. Its sloping roof and two sides are covered in solar panels.[8] The building generates more energy (240,000 kWh per year) than it consumes. In the

FIGURE 4.3 Solar panels on the roof of a house. Dreamstime.com stock house

FIGURE 4.4 Solar panel on a telephone pole. © Sonnenbergshots | Dreamstime.com

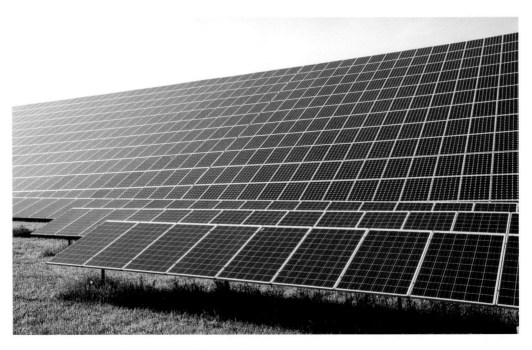

FIGURE 4.5 Solar panels on farmland. Dreamstime.com

FIGURE 4.6 Car park roof with solar panels. Creative Commons.

new city, all buildings could be linked to facilitate the sharing of energy. Such an approach is already being tried in many places. For example, Monash University in Melbourne, Australia, a community of about fifty thousand people, has decided to establish energy sharing among all its buildings. Many of them generate a significant amount of power through PVs. Communities that already share their energy generated by solar cells are Babcock Ranch in Florida (see fig. 4.7) and Pal Town ("solar city") in Japan.[9]

Special energy towers clad in PVs could be distributed throughout the new city and function as energy generation and storage hubs, adding another layer of resilience to the city's infrastructure. In a few exceptions, energy towers might supply electricity

FIGURE 4.7 Babcock Ranch, Florida. All 19,500 homes are powered by solar panels. Courtesy Lisa Hall.

to some buildings whose design excludes the possibility of significant energy production. This strategy is referred to as a *microgrid*. With enough power-generating towers scattered throughout the built environment, it is still a much more sustainable approach when compared with a centralized grid system where a single power station supplies all the surrounding structures with electricity.

Transparent Photovoltaic Cell

Around the time that solar panels began appearing on the roofs of suburban homes, a few research groups began working on the concept of a transparent PV (or TPV), which could be installed in existing buildings.[10] They focused on windows as a potential generator of energy, working on how to convert a clear piece of glass into a TPV and have it still retain its transparency. A few approaches succeeded, resulting in yet another industry.

One variety of TPV is made by applying a specially formulated solution to ordinary window glass. The liquid dries to an impervious weatherproof film. The treated window is still transparent enough to allow for normal visibility, but it can now capture the energy in the ultraviolet and infrared spectrum to generate electricity (see fig. 4.8).[11] Photons dislodge electrons on the window's surface that are then are shunted to the edges of the window where an electrode picks them up and distributes them to the energy network of the building.

The most efficient experimental TPV developed so far converts 13 percent of the sunlight it receives into electricity, more than enough to encourage commercialization of the concept.[12] Low-cost high-efficiency TPVs are still a few years away, but researchers are optimistic. A few buildings have already incorporated them into their design. Within the next ten years, multiple approaches to the production of affordable TPVs should enable an entire city to abandon the use of fossil fuels.

Another approach to the production of TPV uses clear, thin, flexible plastic films into which are incorporated organic molecules with properties selected for their ability to allow photons to dislodge electrons (organic photovoltaic cell or OPV). Efficiencies with laboratory versions as high as 10 percent have been reported.[13] None are yet commercially available.

Transparent solar blinds have also come on the market. Each large window-sized blind can produce enough electricity to power three full-sized laptops. A *Solar* magazine article, "Transparent Solar Panels: Reforming the Future Energy Supply" (February 29, 2020), speculated on the future application of transparent solar panels:

> According to Richard Lunt, the Johansen Crosby Endowed Associate Professor of Chemical Engineering and Materials

TRANSPARENT PHOTOVOLTAIC CELL

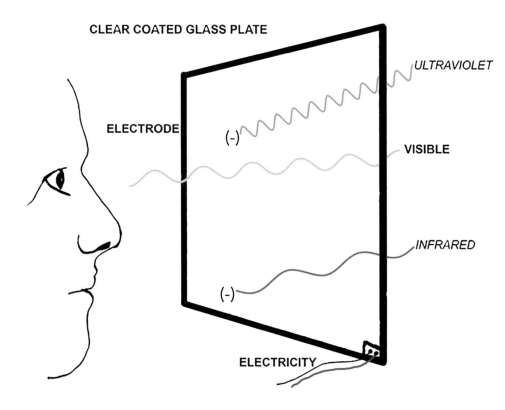

FIGURE 4.8 A special coating on a plain glass window allows visible light to pass through while capturing radiant energy in the ultraviolet and infrared spectra that is then converted into electricity. Drawing by author & CHH.

Science at MSU, highly transparent solar cells represent the "wave of the future" for new solar panel technologies.

Lunt says that these clear solar panels have a similar power-generation potential as rooftop solar, along with additional applications to improve the efficiency of buildings, cars and mobile devices. Lunt and his team estimate that the U.S.

alone has about 5 to 7 billion square meters of glass surface at present.

With this much of glass surface to cover, transparent solar panel technology has the potential to meet about 40 percent of the country's annual energy demand. This potential is nearly the same as that of rooftop solar. When both these technologies are deployed complimentarily, it could help meet nearly 100 percent of the U.S. electricity needs if we also improve energy storage.

The new city could take full advantage of the photovoltaic industry to make nearly every building energy independent.

Wind Power

The amount of wind—a product of the sun aided by the rotation of the Earth—is determined by the combination of atmospheric heating and geographic location (see fig. 4.9). When hot air rises, cooler air flows in to replace it, creating wind in the process. Many countries use wind power to reduce their dependency on fossil fuels (see table 4.1).

In 2022, solar and wind power accounted for a record 28 percent of the total energy consumed in the United States. In most geographic locations winds are seasonal, thereby limiting their usefulness in powering up a city. But in some places winds exceed thirteen miles per hour routinely throughout the year, making them ideal locations for wind turbines.

Early versions of wind turbines had white-colored short blades that turned rapidly, endangering migrating birds.[14] Recent advances in their design removed that design flaw. Blades have become "supersized," and do not need to rotate rapidly to produce the same amount of power. More efficient generators work in conjunction with the newer blade configurations to allow even more energy per turn to be produced, and slower blade rotation gives birds a better chance of avoiding them. Studies have shown

ONSHORE & OFFSHORE WIND RESOURCE MAP

WIND POWER DENSITY POTENTIAL

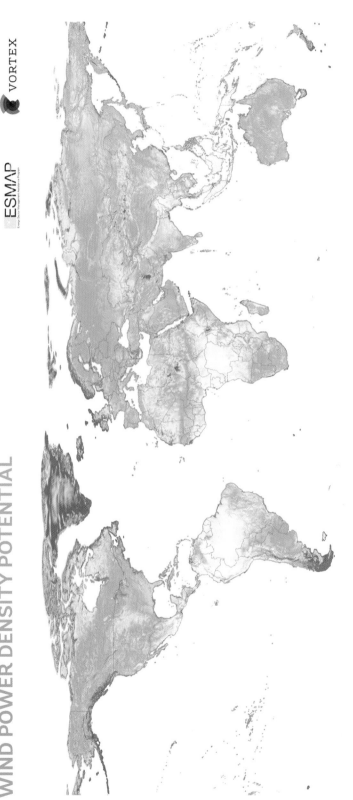

Wind Power Density @ 100m – [W/m]

| <25 | 50 | 75 | 100 | 125 | 150 | 175 | 200 | 225 | 250 | 275 | 300 | 325 | 350 | 375 | 400 | 450 | 500 | 550 | 600 | 650 | 700 | 750 | 800 | 850 | 900 | 1000 | 1100 | 1200 | 1300 | >1300 |

This map is published by the World Bank Group, funded by ESMAP, and prepared by DTU and Vortex. For more information and terms of use, please visit http://globalwindatlas.info

FIGURE 4.9 Global map of wind speed. globalwindatlas.info/en

TABLE 4.1.

Top ten countries by wind power use

Country	2019
China	342 GW
United States	139 GW
Germany	64 GW
India	42 GW
Spain	29 GW
United Kingdom	26 GW
Brazil	19.1 GW
France	18.7 GW
Canada	14.4 GW
Italy	12.7 GW

Source: Jack Unwin and Matt Farmer, "The Top 10 Countries with the Largest Wind Energy Capacity in 2021," https://www .power-technology.com/features/wind-energy-by-country/.

that birds will fly around wind turbines that have a single black blade.[15] Perhaps in the near future, all wind turbines will adopt this bird-friendly feature and eliminate forever the concerns of millions of naturalists.

Wind turbines within the city limits are a rare sight and for good reason. Most tall buildings deflect and slow down wind velocity, lessening their impact on the built environment. In contrast, open plains, ridgelines, deep valleys between mountain ranges, and regions at or near ocean coastal zones offer many opportunities for establishing a steady flow of electricity generated by wind turbines.

Modified wind turbines atop tall buildings have been tried with limited success. Buildings have also been fitted with specially designed horizontal wind turbines (see fig. 4.10). Disappointingly, no turbine design has yet achieved its predicted

FIGURE 4.10 Horizontal urban wind turbines on the roof of an office building. Dreamstime.com

maximum efficiency within city limits.[16] Wilson concludes in his review, "By all means, power your buildings with wind energy, but do it on a larger scale, remotely, where the turbines can operate in laminar-flow winds and where their vibrations and noise won't affect buildings and building occupants." In time, wind turbines may advance to the point of being compatible with the built environment. Until that day arrives, however, the new city will avoid incorporating this method of producing electricity.

Geothermal Energy

Geothermal energy (GE) is abundant in many places throughout the world (see fig. 4.11). Two main energy technologies are employed to access GE. Steam is generated by GE for producing electricity from steam-driven turbines.[17] These municipal power

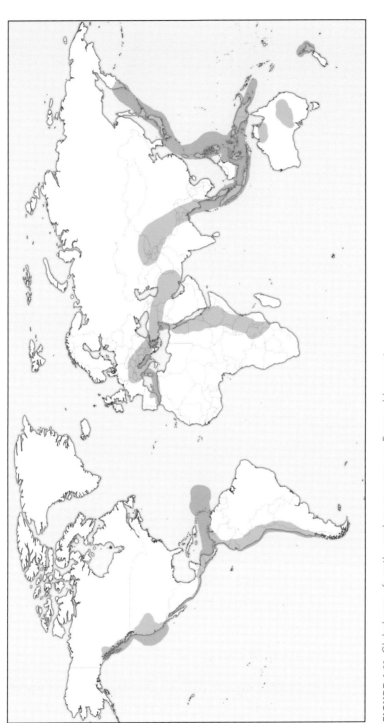

FIGURE 4.11 Global map of geothermal energy sources. Freeworldmap.net

plants are typically part of centralized power grids and are not appropriate for deployment in the new city.

The second application of GE is to drill into deep aquifers of superheated water.[18] The water is cooled to around 85 °F and then piped directly into various structures. For example, Reykjavik, Iceland, heats 87 percent of all its buildings this way. Relying on geothermal energy enables it to avoid producing nearly four million tons of CO_2 each year.[19]

Geothermal energy is an excellent addition to the concept of a gridless economy provided that each building has independent access to the heated water source. Large underground reservoirs of hot water could supply clusters of buildings, another homage to Simard's mother tree and its seedlings.

A new approach to the continuous production of geothermally derived energy has been reported (see fig. 4.12). Cold water from an on-site reservoir is pumped into the ground at

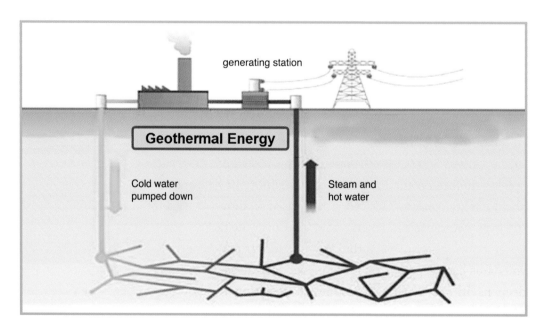

generating station

Geothermal Energy

Cold water
pumped down

Steam and
hot water

FIGURE 4.12 Schematic for continuous generation of electricity from geothermal sources.

the level of dry, hot thermal rock. The water becomes super-heated and is drawn up into aboveground pipes that circulate it through a closed system of coils submerged in a water bath that generates steam. The steam then drives turbines producing electricity. The inflow of cold water is at the same rate as the withdrawal of the heated water, creating an underground current at the level of the geothermal source. The hot water, cooled by giving up its heat to the steam generator system, is returned to the reservoir and can be reused without interruption. Since the Earth is generally equally hot at the same depth wherever the facility is located (except geothermally active sites), this technology could become one of the best answers to any new city's energy needs if it was scalable for use by small clusters of buildings.

Wave and Tidal Power

Two other potential REG technologies include wave and tidal power. Both systems are dependent on a coastal location, limiting where new cities can be established. In addition, only tidal power has received enough attention from commercial firms to become established in a few places. Wave power is still in the experimental phase of development.

One might expect that hydroelectric power could be applied to the new city, but recent severe droughts in the western part of the United States, South America, and elsewhere have cast serious doubts on the long-term sustainability of hydropower as an energy source. Reduced water pressure above dams due to lower water levels in reservoirs slows down the rate at which water flows through the turbines. The Hoover and Glen Canyon dams have experienced lower power outputs as the drought deepens in the American Southwest, resulting in shortages of electricity for entire regions of California, Arizona, and Nevada.[20] It is predicted that this situation will only worsen as RCC increases over the next twenty years, further altering the patterns of global precipitation. Periodic drought also adversely affects the performance of one of South America's largest hydroelectric dams, Itaipu, on the Paraná River, the border between Brazil and Paraguay.

Nuclear Power

Finally, for the sake of completing our discussion of renewable energy resources, we must consider nuclear power. Realizing the energy potential of nuclear fusion and fission reactions has been the dream of physicists since the atom was first split under the football stadium at the University of Chicago on December 2, 1942. Enrico Fermi and colleagues ushered in the nuclear age and the potential for producing nearly unlimited amounts of usable energy that could more than meet the power demands of humanity for millennia to come.

Unfortunately, fission reactions produce large quantities of unwanted radioactive isotopes that need to be safely disposed of. At present, no environmentally acceptable solutions have been agreed upon. In addition, fission reactions are not part of the renewable energy family of technologies. The uranium supply is limited and must be mined and refined, just like other nonrenewable energy sources. Despite these drawbacks, France has adopted nuclear power as its main source of energy generation, producing 75 percent of its electricity this way.

But even if every reactor produced huge quantities of electricity and was operationally safe (e.g., no meltdowns or leaks of isotopes, etc.), situating a nuclear power plant in the new city violates the main reasons for creating a new city: off-the-grid resilience and sustainability.

In contrast, fusion reactions present no danger to the general public. They use small quantities of hydrogen isotopes as fuel. Hydrogen atoms are fused into a donut-shaped plasma called a torus by an array of powerful lasers or a series of ultrapowerful electromagnets known as a tokamak; from that helium is created and with it the release of huge amounts of energy. This same reaction occurs in the sun. Should the fusion machinery ever malfunction, it can be simply turned off and everything comes to a halt. Over the last five years, great strides have been made in fusion energy generation.[21] In the winter of 2022, in one moment, five seconds to be exact, the Culham Centre for Fusion Energy in Oxfordshire, England, operating the Joint European Torus (JET), produced 59 megaJoules of energy.[22] The reaction was halted when the magnets began to overheat. Scientists in charge of the project expected to be able to remedy that problem by first supercooling the magnets.

Many fusion reactors are large objects and therefore must be relegated to the centralized municipal grid paradigm. But smaller ones exist that fit into the basement of medium-sized buildings. These mini tokamaks could be scattered throughout the new city, making fusion energy the all-time winner for renewables. It is possible that someday they will become the next reliable energy source to help eliminate the use of fossil fuels, especially if transparent photovoltaics fail to become more competitive, that is, more efficient.

ENERGY STORAGE SYSTEMS

When solar and wind power become the world's technologies of choice, storing excess electricity using distributed generation (the generation of electricity at or near to where it will be used) will be critical for creating a continuous, reliable flow of energy. Important energy storage systems include

- Mountain gravity (pumped hydro storage)
- Hydrogen (electrolysis of water to hydrogen, stored as a liquid)
- Regenerative fuel cells
- Compressed air storage

- Super magnetic energy storage
- Thermal energy storage (e.g., sand battery)
- Chemical energy storage systems (fuel cells)
- Electrochemical energy storage systems (batteries)
- Hybrid energy storage systems
- Solar fuels
- Graphite solar storage
- Flywheels[23]

Storage systems also include specially configured lithium-ion batteries, conversion of electricity into heated water that is then stored in hyperinsulated containers and the heating of hyperinsulated containers of sand (sand battery). The hot sand is used to heat water for use in warming houses. Generating hydrogen from water by electrolysis, then liquefying it and storing it in underground containers. Hydrogen could also be used in fuel cells to power vehicles and buildings. The water produced by fuel cells would be added back to the electrolysis units for recycling in keeping with the new city's circular water economy. Compressed air is already in use for storing energy, and when released it drives a turbine, generating electricity. Large capacity nickel-lithium storage batteries are commercially available but are currently too expensive for general use.

Flywheel technologies offer another viable option to the mix of energy storage strategies.[24] For example, a spinning flywheel is kept in high rotational mode (stored energy) during the day by input from a building's TPV energy network. At night, the energy could then be accessed to maintain temperature and humidity levels, safety lights in stairwells, and keep surveillance equipment active when the building (for example, an office or a school) is not in use.

The change in our thinking and behavior to create a more balanced, healthier world starts with the switch to renewable energy resources. Powering up the new city with clean, sustainable energy technologies will revolutionize our relationship with the natural world. Eliminating various kinds of pollution that non-renewable carbon-based fuels generate will be the biggest positive change. Petroleum products (i.e., plastics, cosmetics, etc.) will likely still be in demand over the next hundred years, so it is also necessary to improve the efficiency and lower the expense of recycling plastics and other synthetics made from oil. Natural gas will likely be replaced by solar, wind, and hydrogen fuel cell technologies. When that happens, the climate will get back on track and slow its rate of change. Over time, the air will become cleaner and everything and everyone will breathe more easily. As a result, we can expect to enjoy longer, healthier lives.

5

The New City Imagined

I did then what I knew how to do.
Now that I know better, I do better.

—MAYA ANGELOU

Four radical changes to the way cities function—carbon capture, water harvesting from the air, urban agriculture, and
renewable energy strategies, all inspired by nature—if successfully implemented, have the promise of significantly reducing the rate of climate change, restoring damaged terrestrial and
aquatic ecosystems, and improving everyone's health. Clean air,
bountiful and affordable produce; spacious and appealing housing and work environments, and abundant safe drinking water
all favor a more holistic lifestyle. A city that sequesters carbon
into its infrastructure allows everyone to try to reduce our environmental footprint with little effort. Simply by agreeing to live
there, they become part of the solution. Civic pride seems to be
in short supply nowadays, but the new city will hopefully inspire
those who want to help repair natural systems while improving
their own lives.

The absence of private vehicles within the city limits will give
way to reliable, rapid public transportation. Cars for rent will
still be available for those who wish to take an occasional trip
to the verdant countryside. I also imagine that all city transport

systems will be underground and electric or hydrogen fuel cell powered.

The physical surroundings of where one lives and grows up create emotional attachments. That essential niche and social context is set in motion the moment they are born. First discovered by the renowned Austrian behaviorist Konrad Lorenz, imprinting is now recognized as a fundamental principle of the science of psychology.[1] In 1935, Lorenz conducted experiments on various animals to determine when they "attach" to their parents. His most famous experiments involved geese. To his surprise, imprinting occurred the moment the newly hatched goslings detected motion. They would latch onto any moving object and adopt it as their "parent." If, at the moment they emerged from their shells, the young birds heard and saw a human honking away in front of them, in this case, Lorenz himself, then for the rest of their lives they would follow him anywhere he went. More remarkably, he went on to show that the geese were hard-wired to behave that way. For his efforts, in 1973, Lorenz was awarded the Nobel Prize in Physiology.

In addition to biological imprinting, something else occurs after that initial moment, which, in part, determines who we are. It begins to shape us as we become aware of where we are and is reinforced throughout childhood into adolescence. The highly acclaimed Pulitzer Prize–winning book *So Human an Animal* by René Dubos lays the groundwork for the concept of social imprinting.[2] It is a worthwhile read for those not familiar with his global assessment of the future of humanity. I had the pleasure and privilege of spending time with him during my three years as a postdoctoral fellow at The Rockefeller University in New York, where he worked. Dubos argues that people who live in so-called "privileged" neighborhoods behave in ways that support the values and attitudes of those living there, maintaining the social structure of that community. Dubos further extends his observations and insights, concluding that the same is also true for those who live in barrios and slums with deteriorating infrastructure and fewer municipal services than are found in many middle-class communities. That's why he says it is often challenging to bring about change to places that most need it.

But what if an entire city was designed to meet the needs of people solely as they relate to infrastructure and municipal services—a government that recognizes, supports, and respects the lives of everyone who lives there, regardless of age, ethnicity, religious beliefs, or politics? Generating a broad base of community spirit and camaraderie is challenging when municipal services are designed to exclude the poor, creating generations of disenfranchised communities.

I have no interest in promoting yet another utopian society in which a restrictive philosophy forces everyone to behave within a rigid set of rules or risk expulsion. Every effort to establish one of these bizarre collectives has eventually collapsed through rebellion of the oppressed. We are not biologically wired to live in such a world. The worst feature of all failed utopian societies is their obsession with suppressing individuality, restricting creativity, and thus dooming the effort to confine the human spirit in a contrived scheme.

If the new design of an urban environment is enough to encourage the acceptance and sustainability of its institutions, then why not start with that premise? Create human habitats that are both beautiful to live in and whose essential municipal services are largely taken care of by artificial intelligence–driven smart installations: a truly smart city. Oversight for maintenance would be in the hands of teams with expertise in AI management, together with city planners and designers.

Most importantly, the four pillars of sustainability could result in a more resilient infrastructure that by design incorporates collaboration and sharing, emulating in concept and function the mechanisms that unite the trees in a temperate zone forest. By increasing the city's resilience, it will have greater adaptability to environmental change. Using artificial intelligence will create an urban network where buildings share energy. We can extend this concept to include water and food in the shared economy since these two essential resources will be widely available throughout the new city. An abundance of green markets will guarantee that everyone gets the produce they need for their own food needs. Belowground reservoirs of clean, fresh water, generated by harvesting moisture from the air and recycling gray water, could forever change our concept of urban sustainability.

The majority of new cities will most likely emerge slowly by replacing old, obsolete buildings with new construction. This process has been ongoing since cities first came into existence. Building whole new sections at a time will accelerate that process and facilitate the incorporation of linked services. As more of a city is brought up to speed, the financial savings will allow additional portions of the city to convert to this new mode of operation. Compared with a modern city today, the amount of money that could be saved in becoming an autonomous entity is significant. For example, New York spends three billion dollars annually just on solid and liquid waste management. Bringing drinking water to the city adds another five hundred million. The money saved by eliminating those three services alone could help provide funding for essential city services like law enforcement, public health, education, and maintenance of parks.

While the process of replacing outdated buildings is ongoing, not every building needs to be discarded simply for the sake of modernity. There are always a few iconic structures worth saving. They could survive by being retrofitted with renewable energy strategies, water harvesting devices, indoor food production facilities, and waste management infrastructure, which would allow each to participate in the revitalized shared economy.

What I have suggested has yet to happen on a large scale. All the parts are in place and operational, however, ready to be integrated: first into a single building, then into a community, and finally into an urban center of excellence. I hope this will happen soon since our world is in dire need of some good news.

I am convinced that rapid climate change will gradually slow when more renewable energy sources replace fossil fuels. But in the near future, we are in for some tough economic times.

I still have not been able to fully envision what the new city will actually look like. Those who choose to be part of establishing it will determine what form it will manifest. Just as I could not predict the phenomenal growth and acceptance of vertical farming, I hope that this built environment evolves into a nature-friendly, longevity-promoting human institution. To endure over the next millennia, I believe that the four pillars of urban sustainability must play a major role in creating the foundation for the next stage in our evolution: a world free of pollution, hunger, and social injustice.

A final look at the cityscape seems in order. For me, nothing is more exhilarating and entertaining than flying at night. You sit comfortably next to the window staring out at the evolving scene below as everything shifts to night mode. Seen from thirty-six thousand feet on a cloudless night, urban landscapes transform into checkered patterns of brightly colored (mostly yellow) light. Some passengers may ponder just how much energy is needed to keep all those cities lit up, myself included.

Some cityscapes are very recognizable from the air. One of my favorites is Chicago (see fig. 5.1). Its shallow crescent-shaped border with Lake Michigan is unmistakable, and its streets are studies in plane geometry, a vision of city planning dictated in part by nature, exhibiting a degree of regularity that typifies the utilitarian early nineteenth-century wave of urbanization that swept through the American Midwest. Further east, Manhattan (see fig. 5.2) also shows off its grid of avenues and streets, which were laid down shortly after the first reliable water supply filled up its Central Park reservoir in 1837. A closer look reveals the influence of the early Dutch settlement in lower Manhattan below Fourteenth Street, whose streets still echo the path of old dirt roads of New Amsterdam the early colonists established for their horse-drawn carts in the 1600s.

FIGURE 5.1 Chicago as seen at night from the International Space Station. Courtesy NASA.

FIGURE 5.2 Greater metropolitan area of New York City, including Manhattan, as seen at night from the International Space Station. Courtesy NASA.

FIGURE 5.3 Italy as seen at night from the International Space Station. Courtesy NASA.

From the International Space Station, the view at night is nothing short of spectacular. Italy's boot and heel, Palermo, Naples, Rome, Milan, and Venice (see fig. 5.3), clearly discernable, glide past its viewing port at 17,150 miles per hour. Photographing anything while moving at that velocity is tricky, but some clever astronauts have managed to get remarkably crisp images of some of the world's most iconic urban centers. Figures 5.4, 5.5, 5.6, 5.7, 5.8, 5.9, 5.10, and 5.11 show a small selection.

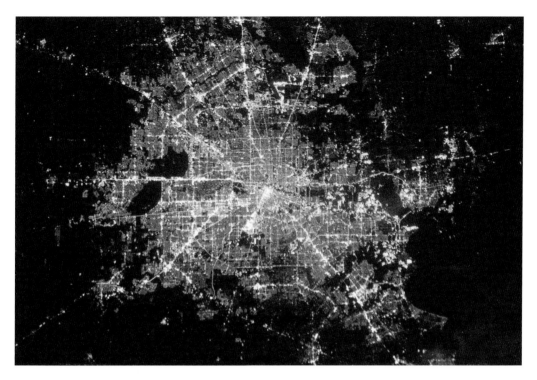

FIGURE 5.4 Houston as seen at night from the International Space Station. Courtesy NASA.

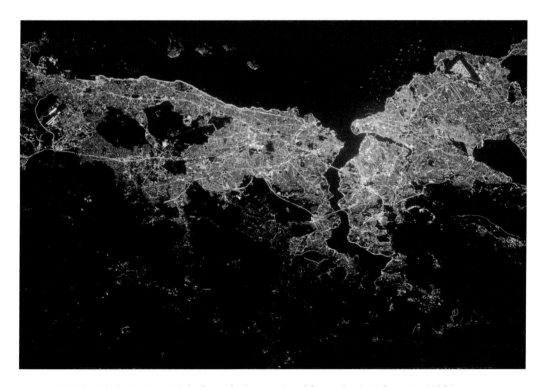

FIGURE 5.5 Istanbul as seen at night from the International Space Station. Courtesy NASA.

FIGURE 5.6 Beijing as seen at night from the International Space Station. Courtesy NASA.

FIGURE 5.7 Berlin as seen at night from the International Space Station. Courtesy NASA.

FIGURE 5.8 Rio de Janeiro as seen at night from the International Space Station. Courtesy NASA.

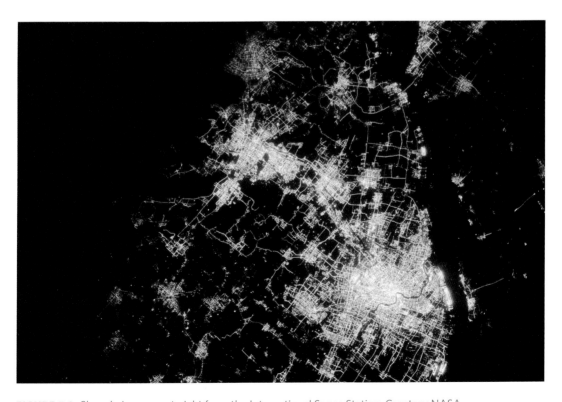

FIGURE 5.9 Shanghai as seen at night from the International Space Station. Courtesy NASA.

FIGURE 5.10 Montreal as seen at night from the International Space Station. Courtesy NASA.

FIGURE 5.11 London as seen at night from the International Space Station. Courtesy NASA.

When viewed together, each city one is unique, they share familiar patterns, especially the way their roads converge at the center of town, regardless of geography. Commerce still rules.

But trips do not last forever. Planes land and astronauts come back down to Earth. Seen up close and personal in the light of day, cities show off all their scars and blemishes, despite our wish for a perfect world. As we walk down a city's back roads and alleys, some of its vulnerabilities are revealed. Trash overflows from half-rusted containers, a pile of discarded, dysfunctional furniture lies next to a building, and assorted animals—feral cats and dogs, rats, pigs, assorted cattle, goats, and chickens (see figs. 5.12, 5.13, 5.14, 5.15, 5.16)—fill the scene, spoiling any romantic illusion we might have held based on its night-time facade of mystery and beauty. In some overcrowded urban centers, all this back-street congestion is in full view along their main thoroughfares. Despite, or perhaps because of all the confusion of daily street life, we are drawn to cities out of irrepressible curiosity. We want to be part of the tumult, an urban rush.

But even up close not all is filth and disorder. A newly erected skyscraper with oddly shaped exteriors and all-glass cladding (see fig. 5.17) instantly becomes a major tourist attraction. Broad tree-lined avenues, eclectic museums, a good mix of international restaurants, beautifully planned and maintained parks, and abundant shopping venues contribute to why we choose city life as our own.

Humanity has the resources, the spirit, and the intelligence to survive, even to thrive. But in order to secure long-term sustainability for our species the world at large must come to terms with today's built environment, correcting its faults while celebrating its myriad positive attributes. Failure to do so will inevitably relegate *Homo sapiens sapiens* to the fossil record. By abandoning the current urban paradigm of the city as a

FIGURE 5.12 Cattle in the street. Photo by author.

FIGURE 5.13 City goat. Photo by author.

FIGURE 5.14 Uncollected garbage. Photo by author.

FIGURE 5.15 Feral dog. Photo by author.

FIGURE 5.16 Hungry cat. Photo by author.

FIGURE 5.17 The IAC building, New York. Frank Gehry, architect. © Sean Pavone | Dreamstime.com

parasite on the surrounding landscape and instead adopting ethical behavior that promotes the concepts of cooperation and respect for all living things, we can achieve balance for all life on Earth.

Appendix

These students at Fordham University, Lincoln Center campus, participated in the vision of Fordhamopolis.

FORDHAMOPOLIS I

Samantha Cassidy

Elizabeth Davis

Olivia Greenspan

Alysa Melendez

Eavan Schmidt

Leigh Anne Statuto

Brian Tang

FORDHAMOPOLIS II

Arica McCarthy

Casey Gritter

Gina Rustami

Anna Creatura

Sam Pajonas

Sophie Cook

Caitlen Perniciaro

Abigail Velasquez

Catherine Priolet

FORDHAMOPOLIS III

Amy Carrillo

Brooke Parret

Cloe Jaquenoud

Cosi Balletti-Thomas

Kellen Stanner

Maddie Griffith

Mobeen Ahmed

Nicholas Rapillo

Renya Wang

Tillie O'Reilly

FORDHAMOPOLIS IV

Hailey Arango

Natalie Bartfay

Julia Duljas

Miles Frank

Val Gorostiaga

Brenda Quach

Vanessa Ryan

Elizabeth Scacifero

Shanen Seale

FORDHAMOPOLIS V

Daisey Bewley

Caleb De Los Santos

Matthew Dyczek

Laura Foley

Grace Getman

Stephanie Lawlor

Danielle Richardson

FORDHAMOPOLIS VI

Spencer Everett

Katherine Hayes

Olivia Johnson

Franchesca Macalintal

Aarushi Mohamn

Maya Reddy

Julia Reynolds

Odalys Tepi

FORDHAMOPOLIS VII

Ryan Chen

Kelly Jackson

Amelia Medved

Jennifer Newton

Shamia Rahman

Genesis Yi

Laurel Applegate

Katelyn Figueroa

Anthony Lekakis

Grace Nelson

Isabela Pizarro

Samantha Roberts

Notes

INTRODUCTION

1. United Nations Department of Economic and Social Affairs, "68 Percent of the World Population Projected to Live in Urban Areas by 2050, Says UN," UN DESA News, May 16, 2018, https://www.un.org/development/desa/en/news/population/2018-revision-of-world-urbanization-prospects.html.
2. World Health Organization, "Urban Planning Crucial for Better Public Health in Cities," WHO Newsroom, May 21, 2020, https://www.who.int/news-room/feature-stories/detail/urban-planning-crucial-for-better-public-health-in-cities.
3. World Health Organization, Road Safety, Global Health Observatory (database), n.d., https://www.who.int/data/gho/data/themes/road-safety.
4. World Bank, *The High Toll of Traffic Injuries: Unacceptable and Preventable* (Washington, DC: World Bank, 2017), https://openknowledge.worldbank.org/bitstream/handle/10986/29129/HighTollofTrafficInjuries.pdf.
5. United Nations, "Generating Power," Climate Action, n.d., https://www.un.org/en/climatechange/climate-solutions/cities

-pollution; City Carbon Footprints, "Global Gridded Model of Carbon Footprints (GGMCF)," n.d., https://citycarbonfootprints .info/.

6. Yale Environment 360, "As the Monsoon and Climate Shift, India Faces Worsening Floods," September 17, 2019, https://e360.yale .edu/features/as-the-monsoon-and-climate-shift-india-faces -worsening-floods.

7. NASA, "Evidence: How Do We Know Climate Change Is Real?" Global Climate Change: Vital Signs of the Planet, updated January 25, 2023, https://climate.nasa.gov/evidence/.

8. Mark Donovan et al., "Local Conditions Magnify Coral Loss After Marine Heatwaves." *Science* 372, no. 6545 (2021): 977–80.

9. NOAA, "Coral Reefs," Office for Coastal Management: Fast Facts, n.d., https://coast.noaa.gov/states/fast-facts/coral-reefs.html.

10. Richard Feely et al., "The Combined Effects of Ocean Acidification, Mixing, and Respiration on pH and Carbonate Saturation in an Urbanized Estuary," *Estuarine, Coast, and Shelf Science* 88 (2010): 442–49.

11. Ashley Bagley, "Sound Chemistry: Ocean Acidification's Effects on Puget Sound," *Currents: A Student Blog* (University of Washington), February 5, 2018, https://smea.uw.edu/currents/sound -chemistry-ocean-acidifications-effects-on-puget-sound.

12. Jane Stephen et al., "Widespread Deoxygenation of Temperate Lakes," *Nature* 594 (June 2021): 66–70.

13. See the Canadian Invasive Species Centre, https://www.invasive speciescentre.ca.

14. Richard Primack and Amanda Gallinat, "Springbud Burst in a Changing Climate," *American Scientist* 104, no. 2 (2016): 102–9.

15. Ryan Shipley et al., "Birds Advancing Lay Dates with Warming Springs Face Greater Risk of Chick Mortality," *Proceedings of the National Academy of Sciences USA* 117 (October 2020): 25590–94.

16. Nathaniel Massey, "Cicadas Swarming U.S. East Coast Are Climate Change Veterans. But Speedier Human-Caused Global Warming May Prove a Challenge," *Scientific American*, May 8, 2013, https://www.scientificamerican.com/article/cicadas-swarming -us-east-coast-are-climate-change-veterans/.

17. City Carbon Footprints, "Global Gridded Model of Carbon Footprints (GGMCF).

18. Thomas Crowther et al., "Mapping Tree Density at a Global Scale." *Nature* 525 (September 2015): 201–5.

19. Richard T. Holmes and Gene Likens, *Hubbard Brook: The Story of a Forest Ecosystem* (New Haven, CT: Yale University Press, 2016).

20. Simard, *Finding The Mother Tree* (New York: Knopf), 283.

21. Ajit Varma, Ram Prasad, and Narendra Tuteja, eds., *Mycorrhiza—Nutrient Uptake, Biocontrol, Ecorestoration*, ed. (Cham: Springer Nature Publisher, 2017); Leho Tedersoo, Mohammad Bahram, and Martin Zobel, "How Mycorrhizal Associations Drive Plant Population and Community Biology," *Science* 367 (February 2020): 867–70 eaba1223.

22. Sarah Batterman et al., "Phosphatase Activity and Nitrogen Fixation Reflect Species Differences, Not Nutrient Trading or Nutrient Balance, Across Tropical Rainforest Trees," Ecology Letters 21 (2018): 1486–95.

1. PILLAR ONE: CARBON STORAGE

1. See Plant for the Planet, www.plant-for-the-planet.org, and its Trillion Tree Campaign, https://www.trilliontreecampaign.org/.

2. United Nations, "Paris Agreement," 2015, https://unfccc.int/sites/default/files/english_paris_agreement.pdf.

3. Frank Lowenstein, Brian Donahue, and David Foster, "Let's Fill Our Cities with Taller, Wooden Buildings," *New York Times*, October 3, 2019, https://www.nytimes.com/2019/10/03/opinion/wood-buildings-architecture-cities.html.

4. Lowenstein, Donahue, and Foster, "Let's Fill Our Cities with Taller, Wooden Buildings."

5. For a good overview of the use of wood in architecture, see https://www.dezeen.com/2021/11/12/dezeen-guide-mass-timber-architecture/.

6. *Nova*, season 44, episode 15, "Secrets of the Forbidden City," directed by Ian Bremner, PBS, aired October 18, 2017.

7. Bhavna Sharma et al., "Engineered Bamboo for Structural Applications." *Construction and Building Materials* 81 (2015): 66–73.

8. Eduardo Souza, "The Potential of Bamboo and Mass Timber for the Construction Industry: An Interview with Pablo van der Lugt," *Archdaily* (blog), November 23, 2021, https://www.archdaily.com/972254/the-potential-of-bamboo-and-mass-timber-for-the-construction-industry-an-interview-with-pablo-van-der-lugt.

9. Gerhard Schickhofer, *Starrer und nachgiebiger Verbund bei geschichteten, flächenhaften Holzstrukturen* (Graz, Austria: Graz University of Technology, 2013 [1994]).

10. Blaine Brownell, "T3 Becomes the First Modern Tall Wood Building in the U.S.," *Architect*, November 8, 2016, https://www.architectmagazine.com/technology/t3-becomes-the-first-modern-tall-wood-building-in-the-u-s_o; Michael Green and Jim Taggart, *Tall Wood Buildings* (Basel: Birkhäuser Architecture, 2017).

11. Thomas Asbeck et al. "Biodiversity Response to Forest Management Intensity, Carbon Stocks and Net Primary Production in Temperate Montane Forests," *Scientific Reports* 11, no. 1625 (2021).

12. The Mother Tree Project, "About Mother Trees in the Forest," https://mothertreeproject.org/about-mother-trees-in-the-forest/.

13. "T3 Becomes the First Modern Tall Wood Building in the U.S."; Green and Taggart, *Tall Wood Buildings*.

14. Green and Taggart, *Tall Wood Buildings*.

15. Tim Smedley, "Could Wooden Buildings Be a Solution to Climate Change?," BBC, July 24, 2019, https://www.bbc.com/future/article/20190717-climate-change-wooden-architecture-concrete-global-warming.

16. Matt Chaban, "Uncanny Valley: The Real Reason There Are No Skyscrapers in the Middle of Manhattan," *Observer*, January 17, 2012, https://observer.com/2012/01/uncanny-valley-the-real-reason-there-are-no-skyscrapers-in-the-middle-of-manhattan/.

17. Ali Amiri et al. "Cities as Carbon Sinks—Classification of Wooden Buildings." *Environmental Research Letters* 15, no. 9 (August 2020).

18. Lowenstein, Donahue, and Foster, "Let's Fill Our Cities with Taller, Wooden Buildings"; Austin Himes and Gwem Busby, "Wood Buildings as a Climate Solution," *Developments in the Built*

Environment 4 (November 2020); Dalina Churkina et al., "Buildings as a Global Carbon Sink," *Nature Sustainability* 3, no. 4 (April 2020): 269–76.

2. PILLAR TWO: URBAN AGRICULTURE

1. Food and Agriculture Organization of the United Nations, "Land Use in Agriculture by the Numbers," May 7, 2020, https://www.fao.org/sustainability/news/detail/en/c/1274219/.

2. Fatimah Mahmood, Muhammad Khokhar, and Zafar Mahmood, "Investigating the Tipping Point of Crop Productivity Induced by Changing Climatic Variables," *Environmental Science of Pollution Research* 28, no. 3 (January 2021): 2923–33.

3. Jared Diamond, *Collapse: How Societies Choose to Fail or Succeed*, rev. ed. (London: Penguin, 2011).

4. Diamond, *Collapse*.

5. Food and Agriculture Organization of the United Nations, "Land Use in Agriculture by the Numbers."

6. Food and Agriculture Organization of the United Nations, "Emissions Due to Agriculture: Global, Regional and Country Trends, 2000–2018," accessed March 8, 2023, https://www.fao.org/3/cb3808en/cb3808en.pdf.

7. Millicent G. Managa et al., "Impact of Transportation, Storage, and Retail Shelf Conditions on Lettuce Quality and Phytonutrients Losses in the Supply Chain," *Food Science and Nutrition* (July 4, 2018), https://doi.org/10.1002/fsn3.685.

8. NASA, "The Causes of Climate Change," accessed March 8, 2023, https://climate.nasa.gov/causes/.

9. Food and Agriculture Organization of the United Nations, "Thinking About the Future of the Food Safety: A Foresight Report," https://www.fao.org/documents/card/en/c/cb8667en.

10. Dickson Despommier, *The Vertical Farm: Feeding the World in the 21st Century*, 10th ed. (New York: Picador, 2021).

11. Precedence Research, "Vertical Farming Market Size to Surpass US$ 31.35 Bn by 2030," *Globe Newswire*, January 12, 2022,

https://www.globenewswire.com/en/news-release/2022
/01/12/2365527/0/en/Vertical-Farming-Market-Size-to-Surpass
-US-31-15-Bn-by-2030.html.

12. Benjamin Durazzo, "Contemporary Applications of Light-Emitting
Diodes in Horticulture: A Review on LED Lighting Technology
and the Use of Wavelength Band and Irradiance Modulation to
Study Plant Photobiology," *Research Gate* (March 25, 2021), https://
www.researchgate.net/publication/350375018_Contemporary
_applications_of_light-emitting_diodes_in_horticulture_A_review
_on_LED_lighting_technology_and_the_use_of_wavelength_band
_and_irradiance_modulation_to_study_plant_photobiology.

13. See the Aldo Leopold Foundation, *A Sand County Almanac*,
https://www.aldoleopold.org/about/aldo-leopold/sand-county
-almanac/ and the Hubbard Brook Ecosystem Study, https://hubbard
brook.org/.

3. PILLAR THREE: HARVESTING WATER FROM THE AIR

1. Harry Lawton and Peter Wilke, "Ancient Agricultural Systems in
Dry Regions of the Old World," in *Agriculture in Semi-Arid Environments*, ed. A. E. Hall, G. H. Cannell, and H. W. Lawton, 1–42 (Berlin: Springer, 1979); Jared Diamond, *Collapse: How Societies Choose
to Fail or Succeed* (London: Penguin, 2011).

2. David Deming, "The Aqueducts and Water Supply of Ancient
Rome," *Ground Water* 58 (October 2019): 153–61.

3. George Rosen, *A History of Public Health* (Baltimore, MD: Johns
Hopkins University Press, 1958).

4. Lewis Mumford, *The City in History: Its Origins, Its Transformations,
and Its Prospects* (New York: Harcourt, 1961).

5. Piers Mitchell, "Human Parasites in the Roman World: Health
Consequences of Conquering an Empire," *Parasitology* 144 (2017):
48–58.

6. Dickson Despommier et al., *Parasitic Diseases*, 7th ed. (New York:
Parasites Without Borders, 2020).

7. Mitchell, "Human Parasites in the Roman World," 48–58.

8. Despommier et al., *Parasitic Diseases*.

9. Filip Havliček and Miroslav Morcinek, "Waste and Pollution in the Ancient Roman Empire," *Journal of Landscape Ecology* 9, no. 3 (2016): 33–45.

10. Rosen, *A History of Public Health*; Fahema Begum, "Mapping Disease: John Snow and Cholera," *Royal College of Surgeons of England* (December 9, 2016).

11. American Society for Microbiology, "Meet the Microbiologist: *Vibrio Cholerae* with Rita Colwell," accessed March 8, 2023, https://asm.org/Podcasts/MTM/Episodes/Vibrio-cholerae-with-Rita-Colwell-MTM-133.

12. "Snow on Cholera, Being a Reprint of Two Papers," JAMA 108, no. 5 (January 1937): 421.

13. American Society for Microbiology, "Meet the Microbiologist: *Vibrio Cholerae* with Rita Colwell."

14. Barbara Shubinksi and Teresa Iacobelli, "Public Health: How the Right Against Hookworm Helped Build a System," Rockefeller Archive Centre, April 30, 2020, https://resource.rockarch.org/story/public-health-how-the-fight-against-hookworm-helped-build-a-system/; Cheryl Elman, Robert McGuire, and Barbara Wittman, "Extending Public Health: The Rockefeller Sanitary Commission and Hookworm in the American South," *American Journal of Public Health* 104, no. 1 (January 2014): 47–58.

15. Guisy Lofrano and Jennette Brown, "Wastewater Management through the Ages: A History of Mankind," *Science of the Total Environment* 408, no. 22 (October 2010): 5254–64.

16. W. MacKenzie et al., "A Massive Outbreak in Milwaukee of Cryptosporidium Infection Transmitted through the Public Water Supply," *New England Journal of Medicine* 331, no. 3 (July 1994): 161–7; https://www.usgs.gov/special-topic/water-science-school/science/surface-water-use-united-states?qt-science_center_objects=0#qt-science_center_objects.

17. USGS, "Surface Water Use in the United States," June 18, 2018, https://www.usgs.gov/special-topic/water-science-school

/science/surface-water-use-united-states?qt-science_center
_objects=0#qt-science_center_objects.

18. Debasish Kundu, Bas van Vliet, and Aarti Gupta, "The Consol-
idation of Deep Tube Well Technology in Safe Drinking Water
Provision: The Case of Arsenic Mitigation in Rural Bangladesh,"
Asian Journal of Technology Innovation 24, no. 2 (2016): 254–73.

19. John Todd and Beth Josephson, "The Design of Living Technologies for Waste Treatment," Ecological Engineering 6 (May 1,
1996): 109–36.

20. Sana Khalid et al., "A Review of Environmental Contamination and
Health Risks Assessment of Wastewater Use for Crop Irrigation
with a Focus on Low and High-Income Countries," *International Journal of Environmental Research and Public Health* 15 (May 2018): 895–90.

21. Iva Hojsak et al., "Arsenic in Rice: A Cause for Concern," *Journal Pediatric Gastroenterology and Nutrition* 60, no. 1 (January 2015):
142–5.

22. Epic Cleantec, "Water Solutions for the Modern World," https://
epiccleantec.com/.

23. Jacques Leslie, "Where Water Is Scarce, Communities Are Turning
to Reusing Wastewater," *Yale Environment 360*, May 2018, https://
e360.yale.edu/features/instead-of-more-dams-communities
-turn-to-reusing-wastewater.

24. Maya Wei-Haas, "The Problem America Has Neglected for Too
Long: Deteriorating Dams," *National Geographic*, May 27, 2020,
https://www.nationalgeographic.com/science/article/problem
-america-neglected-too-long-deteriorating-dams.

25. United States Department of Agriculture, "Drought Impacts on
California Crops," accessed March 8, 2023, https://caclimatehub
.ucdavis.edu/wp-content/uploads/sites/320/2016/03/factsheet3
_crops.pdf.

26. Frank O'Mara, "The Role of Grasslands in Food Security and Climate Change," *Annals of Botany* 110, no. 6 (November 2012): 1263–70.

27. Peter Gleick, *Impacts of California's Ongoing Drought: Hydroelectricity
Generation 2015 Update* (Oakland, CA: Pacific Institute, 2016).

28. A. Park Williams, Benjamin Cook, and Jason Smerdon, "Rapid Intensification of the Emerging Southwestern North American Megadrought in 2020–2021," *Nature Climate Change* 12 (February 2022): 232–34.

29. National Geographic, "From Toilet to Tap," accessed March 8, 2023, https://education.nationalgeographic.org/resource/toilettotap/.

30. Orange County Water District, "Water Reuse," accessed March 8, 2023, https://www.ocwd.com/what-we-do/water-reuse/.

31. See "Appendix G: Selected International Profiles," in the National Water Reuse Action Plan, September 2019, https://www.epa.gov /sites/default/files/2019-09/documents/water-reuse-2019-appendix -g.pdf.

32. Salvatore Pascale, et al., "Increasing Risk of Another Cape Town 'Day Zero' Drought in the 21st Century," *Earth, Atmospheric, and Planetary Sciences* 117, no. 47 (November 2020): 29495–29503.

33. Hennie Pretorius, Solutions for Cape Town Water Crisis. Electricity and Control, July 4, 2018, https://www.crown.co.za/latest -news/electricity-control-latest-news/7414-solutions-for-cape -town-water-crisis.

34. Sarah Fecht, "Study Paves Way for Rainwater Harvesting in Mexico City," *Columbia Climate School*, January 14, 2021, https:// news.climate.columbia.edu/2021/01/14/study-rainwater-harvesting -mexico-city/.

35. Christopher McFadden, "Here Are the Most Advanced Methods to Extract Plentiful Water from Thin Air," *Interesting Engineering*, updated December 10, 2021, https://interestingengineering.com /everything-you-need-to-know-about-air-to-water-devices.

36. Alan Buis, "Study Confirms Climate Models Are Getting Future Warming Projections Right," NASA, January 9, 2020, https:// climate.nasa.gov/news/2943/study-confirms-climate-models -are-getting-future-warming-projections-right/.

37. Government of Bermuda, "Water and Wastewater Master Plan Implementation," July 24, 2020, https://www.gov.bm/articles /water-and-wastewater-master-plan-implementation.

38. UN Climate Technology Centre and Network, "Fog Harvesting," accessed March 8, 2023, https://www.ctc-n.org/technologies/fog -harvesting.

39. Dev Gurera and Bharat Bhushan, "Passive Water Harvesting by Desert Plants and Animals: Lessons from Nature," *Philosophical Transactions of the Royal Society A* 378, no. 2167 (February 2020): 2019.0444.

40. Joanna Knapczyk-Korczak et al., "Improving Water Harvesting Efficiency of Fog Collectors with Electrospun Random and Aligned Polyvinylidene Fluoride (PVDF) Fibers," *Sustainable Materials and Technologies* 25 (September 2020): e00191.

41. Hasila Jarimi and Saffa Riffat, "Review of Sustainable New Methods for Atmospheric Water Harvesting," *International Journal of Low-Carbon Technologies* 15 (2020): 253–76.

42. Paulo Scalize et al., "Use of Condensed Water from Air Conditioning Systems," *Open Engineering* 8, no. 1 (2018): 284–92.

43. Alberti Pistocchi et al., "Can Seawater Desalination Be a Win-Win Fix to Our Water Cycle?" *Water Research* 182 (September 2020): 115906.

4. PILLAR FOUR: RENEWABLE ENERGY

1. United States Department of Energy, "Quadrennial Technology Review: An Assessment of Energy Technologies and Research Opportunities," accessed March 8, 2023, https://www.energy.gov /sites/prod/files/2017/03/f34/qtr-2015-chapter5.pdf.

2. United States Department of Energy (Office of Energy Efficiency & Renewable Energy), "Solar Radiation Basics," accessed March 8, 2023, https://www.energy.gov/eere/solar/solar-radiation-basics.

3. U.S. Energy Information Administration, "What Is U.S. Electricity Generation by Energy Source?" accessed March 8, 2023, https://www.eia.gov/tools/faqs/faq.php?id=427&t=3.

4. National University of Singapore, "New Efficiency Record for Solar Cell Technology," *ScienceDaily*, January 21, 2022, https://www .sciencedaily.com/releases/2022/01/220121124856.htm.

5. Re-volv.org, "History of Solar," accessed April 11, 2023, https://
   re-volv.org/get-involved/education/historysolar/.

6. Luke Richardson, "What is the history of solar energy and
   when were solar panels invented?," Energysage.com, May 3,
   2022, https://news.energysage.com/the-history-and-invention-of
   -solar-panel-technology/

7. Hannah Ritchie, Max Roser, and Pablo Rosado, "Renewable
   Energy," *Our World in Data*, accessed March 8, 2023, https://
   ourworldindata.org/renewable-energy.

8. See "Powerhouse Telemark," https://www.powerhouse.no/en
   /prosjekter/powerhouse-telemark-2/.

9. Mayumi Negishi, " 'Solar City' Proves Allure of Sun's Energy in
   Japan," *Green Business News*, November 11, 2008, https://www
   .reuters.com/article/us-solar-japan-life-btscenes/solar-city
   -proves-allure-of-suns-energy-in-japan-idUSTRE4AA2L
   620081111.

10. Lihui Liu et al., "Toward See-Through Optoelectronics: Trans-
    parent Light-Emitting Diodes and Solar Cells," *Advanced Optical
    Materials* 8, no. 22 (November 2020), https://doi.org/10.1002
    /adom.202001122.

11. RMIT University, "Spray-On Clear Coatings for Cheaper Smart
    Windows," *ScienceDaily*, August 5, 2020, https://www.sciencedaily
    .com/releases/2020/08/200805102018.htm.

12. Kangmin Lee et al., "The Development of Transparent Photo-
    voltaics," *Cell Reports Physical Science* 1, no. 8 (August 26, 2020),
    https://doi.org/10.1016/j.xcrp.2020.100143; "Punching Holes
    in Opaque Solar Cells Turns Them Transparent," *EurekAlert!*
    December 11, 2019, https://www.eurekalert.org/news-releases
    /634370.

13. Qian Kang et al., "A Printable Organic Cathode Interlayer Enables
    over 13 Percent Efficiency for 1-cm^2 Organic Solar Cells," *Joule* 3,
    no. 1 (January 16, 2019): 227–39; Xiaomin Xu et al., "Thermally
    Stable, Highly Efficient, Ultraflexible Organic Photovoltaics,"
    Proceedings of the National Academy of Sciences USA 115, no. 118 (May
    2018): 4589–94.

14. K. Shawn Smallwood, "Estimating Wind Turbine-Caused Bird Mortality," *Journal of Wildlife Management* (December 13, 2010), https://doi.org/10.2193/2007-006.

15. Roel May et al., "Paint It Black: Efficacy of Increased Wind Turbine Rotor Blade Visibility to Reduce Avian Fatalities," *Ecology and Evolution* (July 26, 2020), https://doi.org/10.1002/ece3.6592.

16. Alex Wilson, "The Folly of Building-Integrated Wind," *Environmental Building News* 18, no. 5 (May 2009), https://www.buildinggreen.com/sites/default/files/ebn/EBN_18-5.pdf.

17. Sanya Subir, "Geothermal Power Capacity, Sustainability and Renewability," *Proceedings World Geothermal Congress 2005*, Anatalya, Turkey, April 24–29, 2005.

18. Jon Limberger et al., "Geothermal Energy in Deep Aquifers: A Global Assessment of the Resource Base for Direct Heat Utilization," *Renewable and Sustainable Energy Reviews* 82, part 1 (February 2018): 961–75.

19. Sveinbjorn Bjornsson, *Geothermal Development and Research in Iceland*, ed. Helga Bardadottir (Reykjavik: Gudjon O, 2006).

20. Peter Gleick, *Impacts of California's Ongoing Drought: Hydroelectricity Generation 2015 Update*. Oakland, CA: Pacific Institute, 2016.

21. Peter Dockrill, "A Huge Fusion Experiment in the UK Just Achieved the Much Anticipated 'First Plasma,'" *ScienceAlert*, November 3, 2020, www.sciencealert.com/huge-fusion-experiment-achieves-first-plasma-in-landmark-step-towards-clean-energy.

22. UK Atomic Energy Authority, "Jet Is the World's Largest and Most Advanced Tokamak," accessed March 8, 2023, https://ccfe.ukaea.uk/research/joint-european-torus/.

23. United States Department of Energy (Office of Energy Efficiency & Renewable Energy), "Zero Energy Buildings Resource Hub," accessed March 8, 2023, https://www.energy.gov/eere/buildings/zero-energy-buildings.

24. Sima Koohi-Fayegh and Marc Rosen, "A Review of Energy Storage Types, Applications and Recent Developments," *Journal of Energy Storage* 27 (February 2020): 101047; Mustafa Amiryar and Keith Pullen, "A Review of Flywheel Energy Storage System

Technologies and Their Applications," *Applied Sciences* 7, no. 3 (2017): 286.

5. THE NEW CITY IMAGINED

1. Brian McCabe, "Visual Imprinting in Birds: Behavior, Models, and Neural Mechanisms," *Frontiers in Physiology* 10 (May 2019), https://doi.org/10.3389/fphys.2019.00658; Timothy Johnson, "Imprinting as Social Learning," *Psychology* (January 30, 2020), https://doi.org/10.1093/acrefore/9780190236557.013.649.

2. René Dubos, *So Human an Animal* (New York: Scribner, 1968).

Index

Italicized page numbers indicate figures or tables.

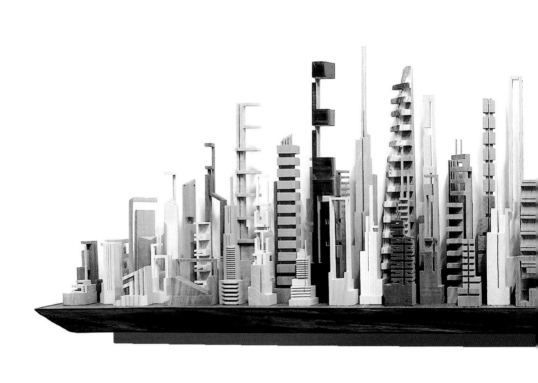

Art by James McNabb